'This fascinating book bursts with the witchy, elusive charms of the plant kingdom's most impudent operators' *... Life*

'... ey's profound scientific knowledge and his passionate fixation with these plants gives one of the most honest and balanced accounts of the plant world ... ever read ... above all, this book is a case for ... of the wild. Read it. You will never look at your ... eeds in the same way again.' *Resurgence*

'... best book on British wild flowers in a decade ... enthralling book' *Independent*

'... ghtly and erudite, this book is an enjoyable romp through the landscape and history of Britain, ... love-hate affair with weeds' *Mail on Sunday*

'... usual mischievous wit, Mabey introduces his ... characters – bindweed, purple loosestrife, poppy, ... the rest – and their various roles as asylum se... rs, opportunists, thugs and flirts ... I would read Mabey on any subject.' *The Times*

'A fascinating display of personal knowledge of the hist... ifferent species and their changing status in ... ds of our ancestors' *Daily Mail*

'... most eye-opening book I have read' Simon Jenkins, *Guardian*

'Mabey's amble through the low-level, high-rise world of weeds is rich in lore and usefulness. As in all his work, what comes over is his abiding passion for plants and the sustenance th... bo... nd spiritually.' *Observer*

RICHARD MABEY is one of our greatest nature writers. He is author of some thirty books including the bestselling plant bible, *Flora Britannica*, *Food for Free* and *Nature Cure* which was shortlisted for the Whitbread, Ondaatje and Ackerley Awards. His biography, *Gilbert White* (Profile) won the Whitbread Biography Award. A regular commentator on the radio and in the national press, he is also a Director of the arts and conservation charity Common Ground and Vice-President of the Open Spaces Society, and he was elected a Fellow of the Royal Society of Literature in 2012. He lives in Norfolk.

## ALSO BY RICHARD MABEY

*Food for Free*
*The Unofficial Countryside*
*The Common Ground*
*The Flowering of Britain*
*Gilbert White*
*Home Country*
*Whistling in the Dark: In Pursuit of the Nightingale*
*Flora Britannica*
*Selected Writing 1974-1999*
*Nature Cure*
*Fencing Paradise*
*Beechcombings*
*A Brush with Nature*

# WEEDS

How vagabond plants
gatecrashed civilisation and
changed the way we think about nature

RICHARD MABEY

P

PROFILE BOOKS

This revised paperback edition published in 2012

First published in Great Britain in 2010 by
PROFILE BOOKS LTD
3A Exmouth House
Pine Street
London EC1R 0JH
*www.profilebooks.com*

1 3 5 7 9 10 8 6 4 2

Designed in Baskerville by Geoff Green Book Design
Typeset by MacGuru Ltd
*info@macguru.org.uk*

Printed and bound by CPI Group (UK)
Ltd, Croydon, CR0 4YY

ISBN 978 1 84668 081 6
eISBN 978 1 84765 284 3

MIX
Paper from
responsible sources
FSC® C020852

To Robin and Rachel

# Contents

I markd the varied colors in flat spreading fields chekerd with closes of different tinted grain like colors in a map the copper tinted colors of clover in blossom . . . the sunny glare of the yellow charlock and the sunset imitation of the scarlet head aches with the blue corn bottles crowding their splendid colors in large sheets over the lands and 'troubling the cornfields' with destroying beauty.

John Clare, 'Leisure', *c*. 1825

# *Acknowledgements*

~

My thanks to: Ronald Blythe for half a lifetime's walks and wisdom, for insights into John Clare and for the gift of the John Nash sketch of woody nightshade (frontispiece). Greg Doran for teaching me much about Shakespeare's use of natural symbols. Chris Fletcher, and the staff of Duke Humphrey's Library, the Bodleian, Oxford for arranging for me to read the Bury St Edmunds herbal. Libby Ingalls for information about American weeds. Molly Mahood for her inspirational book *The Poet as Botanist*, and for allowing me access to her unpublished catalogue of all the mentions of wild plants in John Clare's writings. Leo Mellor for so enthusiastically sharing his work on the literature of Second World War London. Philip Oswald for his elegant translation of Sibthorp's Latin description of rosebay. Jules Pretty for his companionship and knowledge on explorations of the Basildon Plotlands. Jeremy Purseglove for background information about the campaign against Japanese knotweed. Christopher Woodward for stimulating discussions on the aesthetics and literature of ruins. And special thanks to Bob Gibbons for reading the manuscript with an expert botanist's eye when he was meant to be doing something else, and rootling out many foolish slips.

For various tips-off, books and ideas, thanks to Andrew Branson, Clive Chatters, Mark Cocker, John Newton, Martin Sanford and Elizabeth Roy.

For permission to quote an extract of 'Rural Economy', by Edmund Blunden (appears in *Undertones of War*, published by Penguin Books), my thanks to the Edmund Blunden Estate c/o David Higham Associates. Thanks also to Peter Daniels for permission to quote from 'The Shoreditch Orchid'.

My agent Vivien Green was, as usual, a tower of calm support. At Profile, thanks to John Davey and Andrew Franklin for suggesting the book, Penny Daniel for taking a masterful and efficient hold of all the production issues, and Trevor Horwood, who copy-edited with such skill and professionalism. And thanks to Clare Roberts for her exquisite illustrations, which renew a working partnership between us that goes back more than twenty-five years.

Finally, my partner Polly was, as always, a patient and wise helper, not just – so to speak – in the writing room, but in the garden. In both places she knew what needed to be hoed out, and what deserved to stay.

# I

# *Thoroughwort*

## The weed ubiquitous

PLANTS BECOME WEEDS when they obstruct our plans, or our tidy maps of the world. If you have no such plans or maps, they can appear as innocents, without stigma or blame. My own discovery of them was my first close encounter with plants, and they seemed to me like a kind of manna.

I was in my mid-twenties, and working as a publisher's editor in outer London. The job entailed a daily commute from my home in the Chilterns to the urban hinterlands, and I relished the paradox involved in journeying from the sedate order of Home Counties countryside to the wildness of the city. Penguin Books' education division was no *belles lettres* salon, shaded by reflective plane trees. It had been established to pioneer a new kind of textbook, and lay in a defiantly untraditional landscape a mile north of Heathrow airport. This was the Middlesex borderlands, a huge area of wasteland being slowly overtaken by hi-tech industry. Below my office window, the Grand Union Canal wound its flotsam-strewn way towards London, fringed by immigrant plants from three continents. To the west lay a labyrinth of gravel pits, now flooded, and derelict refuse tips whose ancestry went back to Victorian times. They were regularly raked over by bottle collectors, as if they were on the edge of a Third World slum. Northwards our parish frayed into a maze of breakers' yards and trailer parks, where the top predator was the German shepherd guard dog. The whole area was pocked with inexplicable holes and drifts of exotic litter. And most thrillingly to me, it was being overwhelmed by a forest of disreputable plants.

The work I was involved with chiefly concerned developing books on current affairs and social studies for school leavers. 'Relevance' was the fashionable touchstone. The books (more like magazines, really) had what we hoped were accessible but politically challenging texts, and were designed for a readership whose world was in a constant state of edgy flux. When I looked out of the window at the waves of riotous greenery, that world already seemed to be coming our way, fast.

There was nothing pretty or charming about this vegetation, no echo of the wild flowers of the English pastoral – or of England itself, for that matter. But it pulsed with life – raw, cosmopolitan, photosynthetic life. On the tumuli of the old tips, forests of noxious hemlock shot up through the detritus. Indian balsam, smelling of lavatory cleaner but alive with insects, blanketed the thrown-out bottles. Thirty-foot high bushes of buddleia from China towered above the layered sprays of knotweed from Japan, magenta-flowered everlasting-pea from the Mediterranean and the exquisite swan-necked blooms of thornapple, a weed now so spread about the world that its original home is unknown. Beneath them a galaxy of more modest weeds tricked out the compacted layers of plastic and glass that passed for soil. Wormwood, the source of absinthe; three species of nightshade; the horseshoe leaves of coltsfoot; bristly oxtongue, a weed whose scabby leaves looked as if they were afflicted by industrial acne. And strange tuftings that one might see growing wild together nowhere else in Britain except these abandoned places: cumin, feral gourds, fuller's teasel. There was an aura of fantasy about these plants, as if the incantation 'wasteland' made anything possible.

I wandered through this ragged Arcadia in my lunch hours, amazed at its triumphant luxuriance, and feeling, in a naively romantic way, that its regenerative powers echoed the work we were trying to do inside. The plants felt like comrades in arms, vegetable guerrillas that had overcome the dereliction of the industrial age.

This was my entrée into the world of plants, and it has permanently shaped my attitude towards those species usually vilified as weeds. I'm inclined to offer them a second opinion, to wonder what positive features we might glimpse in their florid energy. But I accept that my sixties passion for those Middlesex prodigies was eccentric and probably irresponsible. They were, by most standards, the worst possible kinds of weeds. Many were escapees and trespassers. They had broken out of the disciplined constraints of ornamental gardens and pharmaceutical company farms and were running amok. Several were profoundly toxic. At least two subsequently became so invasive that they're now on a blacklist of species which it's illegal to 'plant or otherwise cause to grow in the wild'. But with weeds context is everything. *Any* plant growing in such shabby surroundings becomes a weed. They're the victims of guilt by association, and seen as sharing the dubious character of the company they keep. If plants sprout through garbage they become a kind of litter themselves. Vegetable trash.

Given the impact of weeds on the planet, it's not always obvious that they are plants whose reputation – and therefore fate – is, in the end, a matter of this kind of personal judgement, that it's in our gift to demonise or accept them. Ever since Genesis decreed 'thorns and thistles' as a long-term punishment for our misbehaviour in the Garden of Eden, weeds have seemed to transcend value judgements, to be ubiquitous and self-evident, as if, like bacteria, they were a biological, not a cultural, category. For thousands of years they've strangled crops and broken backs. In the medieval period they caused outbreaks of mass poisoning, and were given names that suggested they were the Devil's spawn. Today, despite annual chemical drenchings that massively exceed those applied for insect pests, they still reduce arable productivity by 10 to 20 per cent.

And they become more problematic by the year. Across the world global trade has introduced a whole new class of cosmopolitan freeloaders. Striga is a pretty but parasitic snapdragon, whose blossoms in its native Kenya are used to strew across the paths of visiting notables. In 1956 it found its way to the eastern United States, where it has since reduced hundreds of thousands of acres of corn to stubble. Japanese knotweed was introduced to Britain in Victorian times, as an elegant shrub for the woodland garden. In not much more than a century we've become blind to its delicate flower tassels and gracious leaf sprays, and now regard it as the most dangerously invasive plant in the country. The current estimate for clearing it from the

Olympic site in east London is £70 million. None of these outlaw species have changed their identities in graduating as weeds, just their addresses.

Yet even in these two examples the ambivalence and instability of the weed blacklist is clear. The ornamental in one place becomes the malign invader in another. What had been a crop or a medicine, centuries ago, falls from grace and metamorphoses into a forest outlaw. And just as readily the weed is domesticated into a food plant or a children's plaything or a cultural symbol. Mealy leaved fat-hen has been through all these cultural mutations. It migrated from its wild home on the seashore to haunt the middens of Neolithic farmers, from which it was later moved into rough-and-ready cultivation for its oily seeds. Then, as tastes changed, it became a loathed infestation of crops such as sugar beet (to which, ironically, it's related) only to return to partial favour amongst modern foragers.

Of course, 'it all depends what you mean by a weed'. The definition *is* the weed's cultural story. How and why and where we classify plants as undesirable is part of the story of our ceaseless attempts to draw boundaries between nature and culture, wildness and domestication. And how intelligently and generously we draw those lines determines the character of most of the green surfaces of the planet.

~

The best-known and simplest definition is that a weed is 'a plant in the wrong place', that is, a plant growing where you would prefer other plants to grow, or sometimes no plants at all. This works tolerably well, and explains, for example, why English bluebells (whose proper place is the forest) are often weeded out when they spread aggressively inside gardens, while Spanish bluebells (proper place the Mediterranean) are viewed as malignant aliens when they stray *outside* the garden, into the native woodland redoubts of the 'true' bluebell. But there are many nuances of appropriateness and place here, beyond the basic notion of a plant's proper biological home. The sense of a garden as a personal domain is involved; so is a kind of nationalism, even the aesthetic patriotism of seeing in the native bluebell's soft, Celtic curves something more in tune with the British Greenwood than the brasher bells and angular stalks of the Spanish species.

But it's a coarse definition and begs the question of what is the 'right place' for a plant. It would be hard to imagine a more proper location for ash trees than natural, temperate woodland, but foresters call them 'weed trees' when they grow amongst more commercially desirable timber – and, perhaps, because the ash's effortless regenerative power puts in the shade the forester's harder-won achievements. Here, the apparently objective 'proper place' resolves on closer inspection into 'territory', a more personal, culturally determined space.

And the criteria for weediness can change dramatically with time. An early settler in Victoria, Australia, remembered how a fellow Scottish immigrant changed from being a nostalgic reminder of the old country to an outlawed invader: 'One day we came upon a Scottish thistle, growing beside a log, not far from the stable sheds – a chance seed from the horse fodder, of course . . . This was carefully rolled in a piece of newspaper and put under a stone. In a few days it was in a beautifully pressed condition and was shown round with great pride. No one thought that, some twenty years later, the thistle from Scotland would have spread in the new land, and become a nuisance, requiring a special Act in some shires and districts to enforce eradication from private properties.'

Other definitions have stressed other kinds of cultural inappropriateness or disability. Ralph Waldo Emerson opted for usefulness, and said that a weed was simply 'a plant whose virtues have not yet been discovered'. This is a generous and botanically friendly idea, suggesting that reprieves may still be possible for the condemned. But, as with fat-hen, virtues are in the eye of the contemporary beholder. Large numbers of plants were regarded as useful once, only for their virtues to go out of fashion or prove to be bought at great collateral cost. Ground-elder was introduced to Britain by the Romans for the commendable purpose of relieving gout, doubling as a pot-herb into the bargain. But 2,000 years and several medical revolutions later, it's become the most obstinate and detested weed in the nation's flowerbeds.

Toxicity is seen as another ugly and undesirable trait. The most notorious, though far from the most economically damaging weed in the United States is poison ivy, whose impact has been immortalised in a Lieber and Stoller ditty, one of a small group of rock songs to be titled after a weed (Elvis recorded Tony Joe White's 'Poke Salad Annie', for example). In the lyrics, poison ivy is likened to a scheming woman, who'll 'get under your skin', whereupon – and it's one of the great rhyming couplets of pop music – 'You're gonna need an ocean / Of calamine lotion'. In fact calamine can hardly cope with the effects, which are florid and quite out of proportion to what is usually the briefest of encounters. Just the softest brush with a broken leaf can cause nightmarish effects on the skin. It goes red, blisters and itches uncontrollably. If you are susceptible (and fat people are supposedly more so than thin), you can become feverish and oedematous for days. You don't even have to come into contact with the plant itself to catch 'poison ivy' (the effects going under the same name as the plant). You can pick it up from a handshake, or a towel, or by touching the shoes of someone who's been walking in the woods. You can even contract it *indoors*, from the drifting smoke of a bonfire in which there are a few leaves of poison ivy.

By contrast, the British stinging nettle is a minor inconvenience, and deadly nightshade – or dwale, as it's funereally known in some parts – a toxin of not much more than academic interest: at least you have to ingest some part of the plant. Nevertheless, adorned with alluringly jet-black and potentially lethal berries, it's regularly hoicked out of Country Parks and National Trust estates by landowners nervous of litigious visitors. Francis Simpson, the great Suffolk botanist, used to worry that this reflex might threaten an unusual colony of the plant at Old Felixstowe with flowers in an exquisite shade of pale lilac (they are normally a sinister purple): 'There is a danger that one day these plants and their berries may be found by some over-zealous person and destroyed, as frequently occurs with this species. When it is possible I visit the sites and remove the berries, in order to protect the plants.'

Yet in the shadows of this understandable wariness about species that can kill us off a less rational attitude is lurking. Some plants become labelled as weeds because we morally disapprove of their behaviour. Parasites have a bad name because they exploit the nutrients of other plants, regardless of whether they do any real harm in the process. Ivy is vilified as a parasite without even being one. It attaches itself to trees purely for physical support, and takes no nourishment from them. Big tufts can indeed do damage by their sheer physical weight, but the myth of the sap-sucker – the vegetable vampire – is a much more satisfying basis for demonisation.

Simple ugliness or poor posture can also be seen as an infirmity or moral weakness. I can remember the time that small, shy, unathletic children were nicknamed 'weeds' at school; and small, drab, creeping plants such as chickweed and goosegrass can be categorised as weeds for feebleness and a limp wrist as much as for bullying, confirming how plastic and contradictory our definitions are. John Ruskin went so far as to trace out aesthetic and moral standards for flowers. He thought that certain plants were 'unfin-ished' – for example, self-heal, whose flowers and bracts give unsprayed grass a suffusion of purple, like brazed copper, and are hated by lawn obsessives for this very rea-son: 'It is not the normal character of a flower petal,' the arbiter of Victorian aesthetic standards wrote, 'to have a cluster of bristles growing out of the middle of it, nor to be jagged at the edge into the likeness of a fanged fish's jaw, nor to be swollen or pouted into the likeness of a diseased gland in an animal's throat.' Ruskin's disgust echoed the frequent drawing of parallels between human and botani-cal 'savagery'. The nineteenth-century gardening writer J. C. Loudon invited his readers to 'compare plants with men, [to] consider aboriginal species [i.e. wild plants] as mere savages, and botanical species [i.e. cultivars] as civi-lised beings'.

Even wildness itself can be viewed as infra dig when it materialises in the wrong setting. *Helleborus foetidus* (stinking hellebore in English, though this gives an unfair sense of the plant) is a striking denizen of chalky woodland throughout Europe. Its drooping clusters of lemon-green flowers, each tipped with a thin red band, appear as early as February, and shine amongst the dark winter trunks like phosphorus. It's understandably a garden favourite now, but when the distinguished plantswoman Beth Chatto first exhibited it at the Royal Horticultural Society's show in 1975, she was almost disqualified for entering what, because of its wild origins, was classified as a weed.

But the RHS's hauteur is nothing compared to the high puritan criteria applied in Houston, Texas. In that space-age city, by-laws have made illegal 'the existence of weeds, brush, rubbish and all other objectionable, unsightly and unsanitary matter of whatever nature covering or partly covering the surface of any lots or parcels of real estate'. In this litany of dereliction weeds are defined as 'any uncultivated vegetable growth taller than nine inches' – which makes about two-thirds of the entire United States' indigenous flora illegal in a Houston yard. The US Department of Agriculture, struggling to find some unifying principle behind its own pragmatic blacklists, admits that 'over 50 percent of our flora is made up of species that are considered undesirable by some segment of our society'.

We could all make our personal lists on this basis. Mine would include oil-seed rape and cherry laurel. Nothing is sacrosanct when a righteous sense of infestation by the unlovely takes root. I once made a short film with the late Humphrey Brooke, an eminent rosarian who had a sublime garden of some 900 varieties of species and old-fashioned roses in Suffolk. He never pruned his beloved bushes, and hardly weeded round them either. A French journalist remarked of his garden that 'n'est pas une rosarie. C'est un jungle de roses.' But his immense Souvenir de la Malmaison, a progeny of the Empress Josephine's immortal rose garden, produced its heavy, double-cream-coloured, sandalwood-scented blooms deep into winter, and he always sent a bunch for the Queen Mother's Christmas breakfast table. When the filming was over we took the then 70-year-old Humphrey to the local pub, where he got slightly drunk, misbehaved and was ejected. On the way back we passed a suburban garden where the owner was picking modern shrub roses whose shades were a farrago of Day-Glo reds and oranges. Humphrey stopped unsteadily, stared at the scene much as one might at a junk dealer gluing Formica onto a Chippendale table, and screamed 'Vegetable rats!' at the hapless grower.

Weeds are not only plants in the wrong place, but plants which have slipped into the wrong culture.

≈

All these definitions view weeds entirely from a human perspective. They are plants which sabotage human plans. They rob crops of nourishment, ruin the exquisite visions of garden designers, break our codes of appropriate behaviour, make unpleasant and impenetrable hiding places for urban ne'er-do-wells. But is it conceivable they might also have a botanical, or at least ecological, definition? I don't mean by this that they might in some way be close biological relatives: plants tagged as weeds belong to every botanical group from simple algae to rainforest trees. But they have at least one behavioural quality in common. Weeds thrive in the company of humans. They aren't parasites, because they can exist without us, but we are their natural ecological partners, the species alongside which they do best. They relish the things we do to the soil: clearing forests, digging, farming, dumping nutrient-rich rubbish. They flourish in arable fields, battlefields, parking lots, herbaceous borders. They exploit our transport systems, our cooking adventures, our obsession with packaging. Above all they use us when we stir the world up, disrupt its settled patterns. It would be a tautology to say that these days they are found most abundantly where there is most weeding; but that notion ought to make us question whether the weeding encourages the weeds as much as vice versa.

The image of weeds as human familiars is a morally neutral, ecological reflection of the cultural view of them as human stalkers. But they've been companions in a more positive sense. We've had a symbiotic relationship with many of them, a partnership from which we benefit as much as the plants. Because they are common, accessible, comprehensible, weeds were an early port of call whenever some kind of plant material was needed for domestic purposes. Weeds made the first vegetables, the first home medicines, the first dyes. Our ingenuity with them has been boundless. The fronds of horsetail, a persistent weed of badly drained soils and lawns, are covered with tiny crystals of silica. It makes them quite abrasive, and they were once used for polishing pewter and arrow shafts. The piths of soft rush – another invader of compacted soils – were soaked in grease and used as tapers.

Many of the species we've come to call weeds have high

cultural profiles. The common daisy has more than thirty-five local names, and the corn poppy is the one native wild plant whose symbolic meaning is known to everyone. Children, especially, *notice* weeds and revel in their bad reputation and loathsome properties. Wall barley 'flea-darts' (the seedheads stick in the hair) and plantain guns are old games, but instinctively curious children have rapidly discovered the botanical habits of new arrivals too. The explosive pods of Indian balsam, whose seed-hurling abilities are one of the reasons this immigrant species has spread so widely, has now the basis of a highly competitive game, in which kids 'pop' the pods and try to project the seeds as far as possible. (The current record is twelve yards, from the Lake District.) J. K. Rowling understands children's fascination with bizarre plants, and Harry Potter's Hogwarts Academy has an exotic and disgusting weed flora. Bubotuber is a thick, black, slug-like plant, capable of squirming and covered with pus-filled swellings, which cause boils when they touch the skin. Devil's snare winds its tendrils round any hapless creature which gets too close. Interestingly, it can be neutralised by a charm contrived from the bluebell, a 'good' plant, a wild flower, not a weed.

And weeds may have one other benefit. It lurks in our folk-memory, in the practice of fallowing a field between crops, and of composting weeds to cash in on the nutrients they've gathered. My late friend Roger Deakin always used to excuse his failure to weed his vegetable patch by saying 'weeds do keep the roots moist'. Despite their nuisance value to us, weeds may have an ecological point. Their long existence on the planet and all too obvious success suggests that they are highly evolved to 'fit' on the earth in the Darwinian sense, to find their proper niche. Of course they don't have a 'purpose', least of all to deliberately scupper our best-laid plans. Like all living things they just 'are'. But as we survey our long love–hate relationship with them, it may be revealing to ponder where weeds belong in the ecological scheme of things. They seem, even from the most cursory of looks, to have evolved to grow in unsettled earth and damaged landscapes, and that may be a less malign role than we give them credit for.

But in the twenty-first century the spectre has risen of plants that are aggressively weedy in a more fundamental way, species whose reputation is not a matter of personal whim or cultural fashion, botanical thugs that can wreck whole ecosystems as well as human crops and landscapes. The superweed is a favourite villain in science fiction. The seeds of some alien plant-form reach earth, germinate in hours and quickly blanket the planet, or worse, hybridise with humans. A GM crop passes on its herbicide and disease-resistant genes to wild oats, say, and creates the ultimate botanical demon, which perfectly and ironically fulfils the anthropocentric definition of a weed: a rampant plant generated by human activity.

In the real world, the superweed is already here, not as the result of extra-terrestrial invasion but of our own reckless assaults on the natural world. Sometimes a plant is turned into a weed and then into a multinational villain because humans have exterminated all the other wild plants with which it once lived in some sort of equilibrium.

Between 1964 and 1971 the United States sprayed 12 millions tons of Agent Orange on Vietnam. This infamous
mixture of phenoxyacetic herbicides, free dioxins and
turpentine was used as defoliant, to lay bare entire rain-
forests so that the Vietcong had nowhere to hide. It laid
low large numbers of Vietnamese people, too, and is now
banned under the Geneva Convention. But this outlaw-
ing was too late for the forest, which has still not recovered
four decades on. In its place is a tough grass called cogon.
Cogon is a natural component of the ground vegetation of
south-east Asian forest. It flourishes briefly when clearings
are created by falling trees, but retreats when the canopy
closes again. When the trees were permanently obliterated
in Vietnam, it rampaged across the landscape. It is repeatedly burned off, but this seems to encourage it more, and
it has overwhelmed all attempts to overplant it with teak,
pineapple, even the formidable bamboo. Unsurprisingly
it picked up the local tag of 'American weed'. There's some
poetic justice in the fact that cogon recently infiltrated the
USA in the packaging of imported Asian house-plants,
and is now advancing through the southern states.

Other demonic weeds have been created by simple
short-sightedness. In a modern twist of the adage about
a weed being simply a plant in the wrong place, large
numbers of species – potential garden ornaments or food
crops – have been translocated, only to turn into aggressive fifth columnists. They've often been moved thousands
of miles from their native ecosystems, out of reach of all
the nibbling insects and indigenous diseases that usually
keep them in check. Many of these cosmopolitan invaders
come from the fecund sub-tropics, and have a virulence
quite unlike conventional weeds. Australia has been the

hardest hit, with more than 2,500 immigrant species play-
ing havoc with its native wildlife. Globally, these 'invasive
aliens' are regarded as the greatest threat to biological
diversity after climate change and habitat loss.

But even temperate plants can completely change their
character in a new environment. Purple loosestrife is one
of Britain's most beautiful flowers. John Everett Millais
painted its magenta sprays on the riverbank in his picture
of the drowning Ophelia. It is gracious and reserved in its
behaviour, rarely straying beyond the streamside and the
fen. Its name is a literal translation from the Greek, *Lysi-
machia*, meaning deliverer from strife, and Pliny believed
that it was such a powerful promoter of harmony 'that if
placed on the yoke of fractious oxen it will restrain their
quarrelling'. But when it arrived in the New World in the
early 1800s, probably in ships' ballast dug out from European wetlands, it was fated to spark off furious local reactions. The ballast was dumped on the shorelines, and the
loosestrife took root. It came without any of the numer-
ous munching insects that, above and below ground,
keep it in check in Britain, and it took off west like any
other ambitious pioneer. Once established it began moving up watercourses, forming solid stands – miles thick in
places – and pushing native species to the brink of local
extinction. The Hudson River wetlands have been turned
into a solid thicket of purple which even muskrats cannot
penetrate. By 2001 loosestrife had even reached the fragile
marshes of Alaska.

But it's remained a weed of wet places, which is something of a comfort. Given the scale of the diaspora of plant
species, it's surprising that the ultimate plant pest – some

scrambling, fast-growing, leaf-smothering, all-year-round, all-habitat, all-weather Devil's snare – hasn't emerged in reality, and begun overwhelming every kind of vegetation from Amazonian Brazil-nut groves to Hebridean potato plots. The reason it hasn't – and is most unlikely to – is, as we'll see, a profoundly important fact about vegetation, and might help us work out a modus vivendi with the weeds we do have.

~

Organisms 'in the wrong place' are a familiar challenge in the modern world. Beings of all kinds move from one culture to another, creating problems of adjustment on both sides, and sometimes new opportunities, too. Weeds are part of this great company of outsiders, who appear where they are not always welcome. It would be wrong to make glib comparisons between our attitudes towards displaced plants and displaced humans, or to assume, for instance, that an entirely justified concern about invasive plants stems from a kind of botanical xenophobia. Weeds cause trouble in a quite objective sense, and our reactions to and treatment of them are often entirely rational. Nevertheless, the shape of our cultural response to them is familiar. The archetypal weed is the mistrusted intruder. It takes up space and resources that by rights belong to the indigenous inhabitants. Its vulgarity makes it the vegetable equivalent of 'the great unwashed'. Its frequently alien origins and almost always alien ways test the limits of our tolerance. Do we show forbearance and try to accommodate it? Or strive to stop it migrating from its original wild home into our cultivated enclaves? The familiar conundrums of multiculturalism echo in weed ecology, too.

The greatest fears are about the consequences of unplanned integration. The global advance of weed species may be leading towards a more homogenised world, where specialised and local species are driven out by aggressive Jacks-of-all-places, what political scientist Stephen Meyer calls 'adaptive generalists'. 'There will continue to be plenty of life covering the globe,' he writes in *The End of the Wild*. 'Life will just be different: much less diverse, much less exotic, much more predictable, and much less able to capture the awe and wonder of the human spirit. Ecosystems will organise around a human motif, the wild will give way to the predictable, the common, the usual.'

This is happening already. Even by the early twentieth century, many common weeds were virtually cosmopolitan. Bracken, chickweed, knotgrass, curled dock, stinging nettle and bindweed, for example – familiar British natives – now occur on all five continents. The commonest weeds of cities in Europe and North America and Australia are virtually identical. In fact most international weeds were originally of European origin, an ironic side-effect of colonial adventures. But global trade has today put all potential weeds on a more or less equal footing. A list of the top eighteen of 'the world's most serious weeds', compiled in 1977, has just three European plants – fat-hen, field bindweed and wild oats. The bulk of the remainder are aggressive grasses from the tropics, including cogon at number seven, and coco grass at number one, officially recognised as 'the world's worst weed'.

Few parts of the planet are immune. The impeccable

village of Le Fel in the Lot defines itself by its devotion to traditional French culture. The houses are topped with local slates and built on chestnut frames. The surrounding woods are of indigenous tree species managed according to ancient principles. But when I strolled through its narrow lanes in 2008 it felt like a safari through some global botanical park. On the walls and roadsides were naturalised small balsam (from Russia), orange balsam (North America), Indian balsam (the Himalayas), fuchsia (Chile), buddleia (China), Canadian fleabane (North America), Sumatran fleabane (not Sumatra, but South America again) and montbretia (from South African parents via a French plant breeder). The American poet Gary Snyder had a close encounter with botanical aliens while climbing one of the iconic peaks of the American West, Mount Tamalpais: 'We're on a part-trail part-dirt fire road, going through meadows. East into the canyon side, out of the wind, it's deep forest. California Native Plant Society volunteers are along the road wearing Tamalpais Conservation Club T-shirts, rooting out stems and roots. I ask them what, they say, "Thoroughwort, an invasive plant from Mexico."' Thoroughwort is a relative of the asters, and is so-called because the stem appears to push through the leaves. But its name makes it feel like an emblem of the ubiquitousness of modern weeds, which have so comprehensively penetrated our world.

~

However, the weed community shouldn't be judged by the behaviour of its most aggressive members. Weeds – even many intrusive aliens – give something back. They green over the dereliction we have created. They move in to replace more sensitive plants that we have endangered. Their willingness to grow in the most hostile environments – a bombed city, a crack in a wall – means that they insinuate the idea of wild nature into places otherwise quite shorn of it. They are, in this sense, paradoxical. Although they follow and are dependent on human activities, their cussedness and refusal to play by our rules makes them subversive, and the very essence of wildness.

It's this maverick independence that Gerard Manley Hopkins celebrates in his famous couplet 'Oh let them be left, wildness and wet; / Long live the weeds and the wilderness yet' (though his weeds are just commonplace plants of all kinds), and is part of what I will be exploring in this book. The cultural history of weeds is a story of an unresolved paradox that another poet, John Clare, captured perfectly. 'I markd the varied colors in flat spreading fields', he wrote, gazing 'in raptures' over the Northamptonshire wheatfields in which he himself had worked as a weeder, 'chekerd with closes of different tinted grain like colors in a map the copper tinted colors of clover in blossom the sun tand green of the ripening hay the lighter hues of wheat and barley intermixd with the sunny glare of the yellow charlock and the sunset imitation of the scarlet head aches with the blue corn bottles crowding their splendid colors in large sheets over the land and "troubling the cornfields" with destroying beauty'. We have no choice, if we are to survive as a species, to deal with the 'troubling' of weeds. But we can't ignore their beauty either, or their exuberance, or the fact that they are the prototypes of most of the plants that keep us alive. What we ignore, more perilously, is the fact that many of them may be holding the bruised parts of the planet from falling apart.

At one level this book is a case for the defence, an argued suggestion that we look more dispassionately at these outlaw plants, at what they are, how they grow, and the reasons we regard them as trouble. At another level it is a human story. Plants become weeds because people label them as such. For more than 10,000 years farmers, poets, gardeners, scientists and moralists have grappled with the problems and paradoxes they present. It is a huge and ongoing saga, and I have touched on only a fraction of it here, mostly by looking at key moments in weeds' cultural history, when specific challenges by particular species intersected with the obsessions of particular individuals. In the process, I try to explore some of the deeper reasons behind this superficially practical branding we give to such a considerable section of the plant world, and how it reflects our attitudes towards the very idea of nature as an independent realm. The development of cultivation was perhaps the single most crucial event in forming our modern notions of nature. From that point on the natural world could be divided into two conceptually different camps: those organisms contained, managed and bred for the benefit of humans, and those which are 'wild', continuing to live in their own territories on, more or less, their own terms. Weeds occur when this tidy compartmentalisation breaks down. The wild gatecrashes our civilised domains, and the domesticated escapes and runs riot. Weeds vividly demonstrate that natural life – and the course of evolution itself – refuse to be constrained by our cultural concepts. In so doing they make us look closely at the very idea of a divided creation.

# 2

# *Adonis*

The weed before man

ON I MAY 1945, just a week before the VE celebrations
marked the end of the most tumultuous war in earth's history, the director of Kew Gardens gave a talk on a strange
eruption of weeds on London's bomb sites. US troops had
liberated Dachau the day before, but *The Times* – sensing
the lecture's eccentricity, or perhaps some deep metaphor
– covered it as their lead news story. Professor Edward
Salisbury, speaking in the shell of the Savoy Chapel
('itself hit 4 times and damaged 11 times'), had described
how a whole new ecosystem had taken root in the city's
open wounds. It was a story coloured not just by war-
time drama but by the evocative names and addresses of
these vegetable phoenixes, which seemed to suggest more
than accidental links between wild nature and the human
affairs of England's capital city. Bracken carpeted the nave
of St James's in Piccadilly, relishing the dank conditions
created when the bomb-wrecked shell was drenched by
Auxiliary Fire Service hoses. The jazzy chrome flowers of
*Senecio squalidus* (Oxford ragwort – an eighteenth-century
immigrant from the slopes of Mount Etna) had graffit-
ised the rubble of London's Wall. Thornapple, one of the
old apothecaries' most potent nostrums, had sprung up in

the newly sunlit cellars of Cheapside, where it would have been sold to insomniacs and toothache sufferers four centuries before. Gallant-soldier (from Peru), an aptly named if modest posy for the end of a world war, appeared on one in eight of the bomb sites, and the purple surf of rosebay willowherb – already christened 'bombweed' by Londoners – across almost all of them. There were less glamorous familiars, too: creeping buttercup, chickweed, nettle, dock, groundsel, plantains, knotgrass, Genesis's 'thorns and thistles'. Professor Salisbury logged a total of 126 species in all. It was a weed storm, a reminder, if anybody needed one, of how thinly the veneer of civilisation lay over the wilderness.

But there are few indications of how Londoners felt about this invasion of their violated city. Did they see it as a kind of healing, a symbol of the resilience of life in the face of adversity? Or feel that insult was being added to very real injury? It was not, after all, an efflorescence of England's unofficial roses that was exploiting the chaos, but a riot of opportunists and chancers, the spivs of the vegetable world. Perhaps – weeds being perennially ambivalent – they felt both. I doubt they were comforted much by the professor's explanation for the great flowering, that from a weed's point of view, the Blitz was just a good dig-over writ large. If some of them blamed the Germans for the swathes of herbage that braided every patch of roughed-up ground, they were reminded it was their own gardens that many of the seeds had blown in from.

Or was it? It's a frequent refrain in garden writing that weeds are entirely the product of human activities, not just conceptually but physically, as if, by some miraculous bypassing of the normal processes of evolution, they spring up ready formed in the potato patch. 'They cannot survive without us,' insists that usually wise botanical writer Michael Pollan. 'Without man to create crop land and lawns and vacant lots, most weeds would soon vanish. Bindweed, which seems so formidable in the field and garden, can grow nowhere else.' But of course it can, and must. The species which gatecrashed our communities and eventually formed the cultural category of 'weeds' must have had an existence somewhere in the wild from which to begin their imperial expansion.

In 1877 a well was sunk at Meux's Horseshoe Brewery at the southern end of London's Tottenham Court Road, no more than a couple of miles west of what would become the epicentre of the bomb damage seventy years later. It went down 1,146 feet, bottoming out at rocks which were laid down 500 million years ago in the Cambrian era. Much nearer the surface were layers from the Old Stone Age, some 250,000 years ago, when hunter-gatherers roamed what was to become the City of London. And in these layers archaeologists discovered the fossils of a group of plants that were to become not just familiar but, in 1945, briefly famous. Creeping buttercup, chickweed, mare's-tail, dock, knotgrass and a host of other modern weeds had apparently been about in the London basin long before wars or even gardeners were invented. I'm not suggesting that the bomb-site specimens were direct descendants of these archaic plants (though they could have been). But their presence in a landscape entirely unaffected by human activities demonstrates that weed species had a life before, and without, *Homo agricola*.

What is striking, and maybe unexpected, is the similarity

between that Stone Age landscape and the rubble that was the City of London in the 1940s. A quarter of a million years ago the terraces above the Thames were open steppe land, a tremulous rockscape scoured by glaciers, trampled and rootled by herds of mammoth and elk, and flooded every time the ice melted. Any plants that were to survive in these unstable conditions needed to evolve special characteristics. They needed to be adaptable, opportunist and fleet of foot. They needed to be one step ahead of their shifting environment.

It's a generalisation, but a good proportion of the species which were eventually to dog the tracks of humans – crowding into our wheatfields, gardens, building sites, war zones and eventually into the paranoid corners of our imaginations – were those which had made a life in the planet's most restless spaces. They'd evolved on tide-pounded beaches and the precarious slopes of volcanoes, in the flood zones at the edges of rivers and the muddy wallows made by wild grazing animals, in scree and shingle and glacial moraines.

You can still see weed species growing in these aboriginal stations. I've walked the upper reaches of rivers in Yorkshire's limestone dales, where winter floods and rock falls have kept the ground open probably since the melting of the glaciers. All kinds of plants that prefer open habitats muddle together there. Alpine scurvy-grass rubs leaves with seashore-loving thrift, and moorland butterwort grows alongside chalk-down quaking-grass. And amongst them are familiar weeds – coltsfoot, plantain, buttercup, heartsease – also enjoying the openness and opportunity. Of course, there's a good chance the seeds of these particular specimens were washed downriver from the nearest garden, where they'd been enjoying the sun in the borders. But remains of the same species have been found in post-glacial deposits nearby, and this was one of their pre-human homes. Corn poppies glimpsed on Mediterranean shorelines can also have humdrum recent origin, too, from colonies in olive groves and vineyards. But beaches may have been one of their aboriginal homes. And they still crowd amongst black irises high up on the parched and stony hills of the Holy Land, where they were the originals of the New Testament's 'flowers of the field'. (The Mediterranean is the original home of a remarkable number of arable weeds. The region's long dry summers create many patches of bare, arid soil, in which annual weeds can seed and thrive.)

The disturbed borderland between the snow zone and the highest forest is another purely natural weed nursery. In the mountains of northern Greece, the melting snows of late spring sometimes spill out a chaplet of brilliant scarlet. This is pheasant's-eye, *Adonis annua*, a member of the buttercup family that eventually found its way to Britain, mixed up with the seed corn of Neolithic immigrants from the Mediterranean. *Adonis*'s fortunes have been a parable of the changing status of weeds. In the medieval period it was abundant in cornfields, especially in chalky areas. In the sixteenth century the gardener and herbalist John Gerard admired the elegance of its cup-shaped flowers and the black beauty-spot at the base of its petals, and got hold of seeds from the West Country to grow in his garden. He called it 'Rose-a-ruby'. Two hundred years later it was being hawked round the streets of Covent Garden as 'Red Marocco', a fashionable bouquet for the times. But by the end of the nineteenth century new seed-screening techniques had virtually eliminated it from Britain – until 1971, when the warm glacier of an advancing motorway briefly resurrected it in Wiltshire, where the M4 sliced through an area of old cornland. Today it's honoured by being on the UK list of Species of Special Conservation Concern, having come full circle from pest to protected species.

Weeds' filaments of life seem as persistent and pervasive as myths. They survive, entombed in the soil, for centuries. They ride out ice ages, agricultural revolutions, global wars. They mark the tracks of human movements across continents as indelibly as languages. It was partly this indomitability that sparked the imagination of the young Edward Salisbury. He was born in 1886, a product of the Hertfordshire dynasty of Salisburys, and was ferreting about in the local countryside from an early age. When he was still a teenager he showed signs of the curiosity and ingenuity that were to mark out his work as a botanist. He found a plant that he didn't recognise growing on a heap of flints near Harpenden, and sent it to Kew Gardens (where he was later to work) for identification. They informed him that it was a North American species called ragweed, a scruffy member of the daisy family, and the most notorious cause of hay fever in the USA. Salisbury then 'made inquiry', as he put it, which revealed that the flints had been brought from America to England as ballast in a ship. The ragweed seeds (they're armed with spines) must have clung to the flints, survived a transatlantic voyage, and then found the environment of a Hertfordshire roadside a passable substitute for the American brush.

Edward Salisbury went on to study and work at Imperial College in London. He read the works of Charles Darwin, and found that the great biologist's curiosity and unconventional experimental methods chimed with his own. Darwin was fascinated by weeds, as examples of evolution on the fast track. He considered the dispersal of their seeds by sea, and tested the effects of saltwater on germination. He wondered if seeds might travel in the stomachs of dead birds, and sprouted seeds he had extracted from the dung of migratory locusts. He raised more than eighty plants from the mud-ball gathered round a wounded French partridge's leg. His famous 'weed-patch' at Down House in Kent was the first quantified experiment on the competitiveness of weeds. Darwin cleared and dug a plot three feet long by two feet wide, and simply observed what plant life spontaneously emerged: 'I marked all the seedlings of our native weeds as they came up, and out of 357 no less than 295 were destroyed, chiefly by slugs and insects.' This might be a heartening statistic for gardeners, were it not for the sixty-two that were *not* eaten. Darwin does not say what species they were, but they were doubtless familiar adversaries.

Salisbury's own experiments were very much in the Darwinian mould. He wanted to explore those qualities of endurance and mobility that made weeds – especially the traditional weeds of cultivated ground – so successful. The tests he devised are reminiscent of the kind of games children play with plants, and must have needed a special suspension of adult decorum in a man whom colleagues remember as having a taste for high collars and spats. To test the efficiency of airborne dispersal of plants such as thistle and dandelion, whose seeds are kitted out with devices to catch the wind, he stood on a ladder in a draught-free room, dropped the seeds and timed how long they took to fall ten feet. Buddleia's winged fruits took five seconds, groundsel's parachutes eight seconds and coltsfoot's twenty-one seconds. Rosebay's feathery plumes took the best part of a minute to sail placidly to the floor, partially explaining how it had dispersed so widely through London's bomb sites. He combed through animal dung and bird droppings to discover whether these were also agencies for ferrying weeds about – and then planted the excrement-coated seeds in pots to see if they were still fertile. (Sparrows were especially effective carriers when they were common and Salisbury grew plantains, groundsel, chickweed and shepherd's-purse from their droppings.) He even regarded himself as a potential carrier, and famously raised 300 plants of over twenty weed species from the debris in his trouser turn-ups – apparently great gatherers of propagules, despite the spats. 'Since many such seeds are loose in the "turn-up",' he noted fastidiously, 'some from time to time become jerked out so that the wearer becomes a peripatetic censer mechanism, scattering seeds as he walks about.' He repeated the experiment with the mud scraped from his shoes, and found that 'one quite commonly conveys at least six propagules in such a manner'. His findings must have been a shock to those who had never considered themselves parties to such common conveyance.

Weeds often produce seeds in prolific numbers. A good-sized mullein or Canadian fleabane can release in excess of 400,000. Weed seeds have evolved devices to ensure they get ferried to the widest range of new habitats. They can be armed with hooks, burrs, spines, ribs, hairs, to help them stick to passing animals (or botanists' legwear). There is also seed glue. The common garden weed shepherd's-purse is named for its seed heads, which resemble the little pouches or skrips worn by medieval peasants (there's a typical skrip in Brueghel's painting *The Peasant Dance*). Open up a purse and the seeds spill out like tiny gold coins. They're covered with a thin layer of gum, which becomes stickier still when it's moistened – as for instance by contact with the soil – so that it can cling to the feet of birds.

But the survival tool which marks out most weeds of cultivation from other kinds of plant is their relationship with time. To thrive amidst constant disturbance, they need to either grow fast or bide their time. Many weeds have rapid life cycles, or the capacity to lie dormant under the soil for long periods, or both. Tumbleweed seeds can germinate in thirty-six minutes. Groundsel can go through an entire life cycle from seed to flower to seed in just six weeks. In 1765 Gilbert White's Hampshire garden was overrun with a brand-new crop of it as late as October.

As for dormancy, Edward Salisbury had some first-hand experience. His garden in Radlett lay on the site of what had been a cornfield in the Napoleonic wars, and which had reverted to grassland when the price of wheat fell with the peace. When the turf was lifted again to make Salisbury's garden in 1928, patches of an extraordinarily rare cornfield weed appeared. Blue pimpernel, very like a scarlet pimpernel but with cobalt blue flowers, is frequent on the Continent, but began to vanish from England after the agricultural revolution. The seed it emerged from at Radlett must have been buried for more than a century.

Something similar happened in the Chilterns in the

1980s. When the worked-out chalk quarries at Pitstone were being transformed into a nature reserve, the warden Graham Atkins discovered a barn full of topsoil, which had been scraped off the land before it was first quarried in the 1930s. The intention at that time was to fill up the pits once the extraction was finished (probably with rubbish), then spread the stored soil back on the land and return it to agriculture. But different land-use priorities had emerged in the half-century since. The land was to be given back to nature, not farming, and the topsoil was redundant. But Graham Atkins realised that it had been removed before the introduction of chemical weedkillers, and was in all likelihood a living fossil, a huge seed-bank of the exuberant cornfield weeds of an earlier generation of farming. So he spread it on a patch of the reserve instead. The following spring the field burst into bloom, with an explosion of weeds that had not been seen in the area for decades. Blue cornflowers, purple corncockle, yellow corn buttercup. The long, comb-like seed pods of shepherd's-needle. The purple and yellow pagodas of field cow-wheat. And a handful of the long-straw wheat variety they had once grown with.

These few decades of lingering vitality are nothing compared to the centuries that have been recorded for some weed seeds. Dock seeds still germinate freely after sixty years. Fat-hen's have sprouted after being recovered from deep within an archaeological site 1,700 years old. But even these are trumped by the weld (dyer's-rocket) that appeared after the excavation of a Roman site at Cirencester nearly 2,000 years old. Dormancy is an insurance policy, the botanical equivalent of savings put away for a rainy day. If you are a plant species that has evolved to tolerate a turbulent environment, one of the adaptations that will help you survive is to have a percentage of your seeds hold off from germinating for two, three, thirty, three hundred years – just in case the ground isn't roughed up again till then, or the first generation of seedlings are killed. An experiment by two of Salisbury's colleagues on the germination of artificially buried seeds found that after thirty-nine years in the soil, 91 per cent of thornapple seeds would germinate, 83 per cent of black nightshade and 53 per cent of greater bindweed.

Dormancy is still not fully understood. Some species coat their seeds with different thicknesses of rind, or contain water-soluble substances which inhibit germination until they've been leached out by water in the soil. Others seem to respond to warmth, and only sprout when they are in the topmost layers of soil. A few seem to possess an interior clock, a biological chip with a long countdown period.

Corn poppy's dormancy is legendary. Laboratory data (for instance, at least 15 per cent of poppy seeds bide time till germination) simply put numbers to a process that was indelibly engraved in our cultural memory at Flanders Field. Europe's earth is full of poppies and bleeds with them when it's cut. They provide such extravagant evidence of their powers of survival – a field of poppies is visible a mile away, like a grounded sunrise – that they have come to stand as an emblem of the persistence and ambivalent meanings of the entire weed nation. In 2009, as if to prove that as well as immeasurable powers of adaptation they had memories too, an immense throng of late poppies flowered on a Dorset estate in early November, Armistice week.

A single poppy head contains 1,000 seeds, and each plant carries as many as fifty heads. When they are ripe, the head dries out, the roof lifts and a row of tiny holes appears round the rim. The stalk dries out too, and bends under the weight of the seed, scattering them whenever the wind blows, up to three feet away from the parent plant. Of, say, a total production of 20,000 seeds some 85 per cent, or 17,000, will germinate in their first year, if conditions are right. In the second year, maybe another 1,000, and in the third, 500 . . . no one has yet continued an experiment for long enough to witness the far limit of poppy-seed dormancy. But it's been estimated that in the days before chemical weedkillers, an acre of cornfield may have held up to 100 million dormant seeds. Fallowing, weeding, even the temporary occupation of the ground had absolutely no effect. Come the next ploughing, or the next war, a myriad seeds-in-waiting came into their moment, germinated, bloomed and fruited, and scattered a hundred-fold more seeds back into the soil. Poppies must have seemed like an insistence by the earth, the wheat's predestined mate.

No wonder the Assyrians called them 'the daughters of the field', and that the plant's first recorded name – *pa pa* in Sumerian – has remained virtually unaltered for 6,000 years. The Romans regarded poppies as sacred to their crop goddess Ceres. Garlands for her statues were braided from poppy and wheat stalks, and poppy seeds were offered up in rituals to ensure the fertility of the crops. Even in the intensely Christian atmosphere of medieval Britain, when farmworkers were doing what they could to yank out this cussedly beautiful invader, they also paid respects to it. Many of the ancient vernacular names, such as 'thunderflower' and 'lightnings', reflected the superstition that poppies must not be picked, for fear of provoking a storm; and conversely, perhaps, that while they were unpicked their companion crops were protected from downpours. (In Northumberland the chance of lightning striking the picker was greater if the petals fell off 'in the act'; 'nor was the risk small', commented the earnest folklorist G. Johnson, 'for the deciduousness of the petals is almost proverbial'.)

There were early scientific intimations of poppy's adaptability, too. In 1660 the great East Anglian naturalist John Ray noted that 'the seeds of poppy are still capable of germination after ten years', and – two centuries before Darwin – intuited that their diversity in some way helped the poppy's survival. 'The smaller the seeds', he writes in his account of the flora of Cambridgeshire, 'the more fertile they are. Because the smaller the seed the greater number there can be: the seed adapts itself more readily to the climate and so adjusts itself to the environment.'

Twenty years after the publication of Darwin's *The Origin of Species*, the vicar of the small Surrey village of Shirley did a remarkable plant-breeding experiment with poppies which confirmed how the diversity encoded in those myriads of seeds – a variegation in many dimensions beyond dormancy – helped the species' survival. In 1880 the Revd William Wilks discovered, in what he described as 'a wilderness corner of my garden', a patch of poppies in which one solitary scarlet flower had a narrow white edge. He saved the seed and planted it out. The following year, out of some 200 plants, five grew to have flowers on which all the petals were edged with white. The process went on for some years, with the flowers acquiring progressively

more infusions of white, until Wilks had a full set of pale
pinks, and one pure white flower. Then he set himself the
task of changing the mascara-black smear at the base of
the petals to yellow or white. Eventually he had a strain
in which the seeds from any one pod were a genetic pot-
pourri, and would produce petals varying in colour from
deep scarlet to pure white, with all shades of pink between
and unpredictable varieties of white flecking and edging.
He called them 'Shirley poppies' after his village, and they
have become an enduring favourite in cottage gardens. A
correspondent who lives in the road where Wilks bred his
poppies tells me that occasional 'Shirlies' could be found
in the adjoining field until the late 1980s, despite most of
the local farmland having been turned into a golf course.

The Surrey vicar had effectively subjected his poppies
to an accelerated form of evolution by natural selection –
except that he was doing the selecting. He eliminated – as
if he were a discriminating herbivore – all the seedlings
that weren't 'fit' for his purpose. The poppy – fecund,
polymorphous, rapid-cycling – grabbed the opportunity
to get just one seed through this narrow window of opportunity. It is this process which explains the weed paradox –
that weeding encourages the weed. We in effect challenge
the unwanted prodigy to produce forms that slip through
our control systems. It does not take much to beat us. One
seed in a thousand may germinate later than the last hoeing, pass through the sieve intended to exclude it, show
a mysterious immunity to weedkillers. The following year
there are five . . .

~

The poppy is a motif throughout this book. The characteristics that make it a survivor are common to all successful weeds. As a type they are mobile, prolific, genetically diverse. They are unfussy about where they live, adapt quickly to environmental stress, use multiple strategies for getting their own way. It's curious that it took so long for us to realise that the species they most resemble is *us*. Once farming had begun, simultaneously creating the cultural idea of weeds and then setting out to eliminate them, our two orders of creation were irrevocably linked.

# 3

# *Knotgrass*

## Weeds as parable

THERE ARE TWO EXTREME VIEWS of the authors of Genesis – that they were amanuenses for God or politically motivated propagandists. What is inescapable is that, whatever their inspiration, they were preoccupied by plants and their metaphors. They see the world through vegetable allegory and myth. They place 'Grass and herb yielding seed after his kind' in their proper evolutionary position in the Creation story, before the fish and birds and mammals. They stage the great drama of the Fall in a garden. The plot is elaborated largely through botanical symbols – fruit and herbs, contrasting styles of cultivation, a magical and forbidden tree. And the denouement is exile from the carefree life of foraging to the toil of farming and the eternal curse of 'thorns and thistles'. Genesis helped form a moral context for weeds, to stigmatise them as more than a simple physical nuisance. The text itself may even have been partially prompted by the proliferation of weeds in the Middle-Eastern 'cradle of civilisation'.

The first written versions of the Genesis Creation myth (the stories in it are much older) date from about 600 to 500 BCE, and appeared in the region known as Canaan, (modern Mesopotamia), part of the 'Fertile Crescent' where agriculture was first developed. There are two versions of the myth. In the first, God creates humans immediately after the animals, and establishes them in their role as agriculturalists, rulers of the rest of Creation. 'Let us make man in our image, after our likeness,' says God, rather tellingly slipping into the royal 'we'; 'and let them have dominion over the fish of the seas, and over the fowl of the air, and over the cattle, and over all the earth, and over every thing that creepeth upon the earth'. Despite this abundance of subservient and highly edible creatures, God seems to insist on a strictly vegetarian lifestyle: 'I have given you every herb bearing seed, which is upon the face of all the earth, and every tree, in which is the fruit of a tree yielding seed; to you it shall be for meat.'

The second version (Genesis 2 and 3), which introduces the Garden of Eden, is more complex. It takes off at the point when the creation of heaven and earth are complete, and has God creating man in advance of other creatures. He forms him from 'the dust of the ground' and places him in 'a garden eastward in Eden'. It contains 'every tree that is pleasant to the sight, and good for food; the tree of life . . . and the tree of knowledge of good and evil'. Adam's responsibilities are to 'dress it and keep it', to eat what he likes but to stay away from the tree of knowledge. Only then are the animals created and brought before Adam to be named – though his mate, fashioned from one of his ribs, is at this stage called simply 'Woman'.

Then they eat the fruit of the tree of knowledge, and all hell breaks loose. God's punishment is severe and unambiguous. Life will become a vale of tears, ending in death. Women will bear children in pain and sorrow, and become their husband's serfs. And the unbidden harvests of the

Garden will now have to be won by hard toil, *agricultural* toil: 'cursed is the ground for thy sake', God rages, 'in sorrow shalt thou eat of it all the days of thy life; thorns also and thistles shall it bring forth to thee; and thou shalt eat the herb of the field; In the sweat of thy face shalt thou eat bread, till thou return unto the ground.' Then he expels Adam (no mention of Eve) 'from the garden of Eden, to till the ground from whence he was taken'. It was a stark change from those gentle gardener's duties of 'dressing and keeping'.

What is striking in the ecological subtext of Genesis is its sense of bitterness about the arrival of agriculture. Farming here isn't the sacrament of later Western Christianity, in which 'to plough the fields and scatter the good seed on the land' was seen as a metaphor of God's sowing the earth with righteousness. For at least one group of disgruntled Assyrians their farming labour seemed sufficiently cursed by literal and metaphorical weeds to be seen as a punishment or a poisoned chalice, and certainly no substitute for the freedoms of the hunter-gatherer's life.

The sense of loss may have had deep roots. The Garden of Eden is a dramatic device, intended to give the idea of the Fall a tangible power. But the geographical references in Genesis – especially the proximity of Assyria and the Euphrates – suggest that its inspiration was some part of the area known as Mesopotamia, where agriculture had been developed more than 7,000 years before. It's highly doubtful if the idea of cultivation occurred in a sudden 'Eureka!' moment. Most likely it emerged seamlessly from the process of gathering and storing wild food plants. Many scenarios have been dreamed up in the absence of real evidence about what happened. Cultivation was suggested by the rootling of wild animals, which seemed to stir up plant germination. Or by the tendency of some edible wild plants to grow in convenient clumps, which were tended in situ by foragers. Or by the sprouting of harvested plants in the vicinity of settlements. Foragers would tend to choose plants with early-forming or large leaves or seeds, and these characteristics would have been passed on to any plants germinating from food remains at their settlements. The refuse heap may have been the first serendipitous crop field. As the agricultural historian Nikolai Vavilov argued early last century, a wild plant benefited from being selected in this way, and 'seeming to intrude upon the agriculturalist in order to be cultivated . . . seeking shelter near his dwelling, proffering its services'.

The wild plants that Stone Age Mesopotamians gathered included species which today, ironically, are classed as weeds, and which are still used by their descendants in the Euphrates basin. Modern Iraqi villagers gather wild greens from the hills. Mallow species are used in soups and stews. The Mesopotamian salad bowl could be easily assembled in modern Britain, and includes watercress, dock and dandelion. There is, and was, an abundance of wild fruit too: chestnuts, almonds, figs and olives (though these have been gathered and spread about for so long that no one is sure of their natural home). Apricots and pomegranates were also widespread, and if there was a real-world analogue of the fruit that led to Eve's downfall it was probably one of these, since apples don't grow well in the parched climate of the Middle East. (Though Terence McKenna, in his audacious book *Food of the Gods*, makes

a plausible if unsubstantiated case for the tree of knowledge being the hallucinogenic desert mushroom *Stropharia*
(now *Psilocybe*) *cubensis*.) No wonder the local tribespeople
had pangs of nostalgia for their hunter-gathering lifestyle.

What changed their lives – and eventually the whole
course of human civilisation – was the domestication of
a desert weed, a grass called wild emmer. This was gathered from the wild to begin with, and its starch-rich seeds
used in gruels. It has the habit of growing in large clumps,
with the seedheads all at roughly the same height, which
must have been cues in hastening the idea of systematic
harvest. The gatherers – acting in exactly the same spirit
as the Revd Wilks with his poppies – would have prefer-
entially chosen specimens which met their needs: bunches
whose seeds ripened at the same time, and whose hulls
didn't shatter and spill the grain. These traits are genetic,
and would have been passed on to the wild emmer which
sprang from seed spilled near the settlements. These first
steps in domestication were followed by the development
of all the associated agricultural techniques of irrigation,
tilling, communal harvest, threshing, winnowing, mill-
ing, and finally baking. All unrelievedly performed 'in the
sweat of thy face'.

And in that first moment of deliberate cultivation the
concept of 'the weed', the unwanted trespasser, the plant
in the wrong place, was added to the tribulations of the
first farmers. A plot of ground dedicated to cultivation
not only redefined the status of those plants in it that were
*not* cultivated, but physically encouraged them. It was a
field day for any local species that could exploit disturbed
ground. The primeval wheatfields would have been thick
with poppies, black mustard, wild gladioli and tares – not

the pea-family tares of modern Western floras (*Vicia* species), but the toxic grass darnel, which was to haunt European farmers until the late Middle Ages. There was no field-weeding at this stage. The crop and the weeds were crudely separated by hand after harvest – a process faithfully recorded in the parable of the Good Seed. The version in Matthew's Gospel concerns a householder whose fields have been blitzed with weed seeds by an enemy. He advises his workers not to pull them up: 'Nay; lest while ye gather up the tares, ye root up also the wheat with them. Let both grow together until the harvest: and in the time of harvest I will say to the reapers, Gather ye together first the tares, and bind them in bundles and burn them: but gather the wheat into my barn.' This is one of the few weeding techniques which, in the long run, doesn't benefit the weeds. Almost every early agricultural practice inadvertently selected in favour of – and therefore encouraged – botanical fifth columnists, weeds whose form and behaviour most closely mimicked those of the crop they grew amongst. Successful weeds were those species whose seeds were able to smuggle themselves into the seed corn and thus into next year's sowings.

The obstinacy and pervasiveness of weeds must have exasperated early farmers. Yet if, by some technological leap into the future, they had succeeded in controlling them, it's doubtful that what we understand as agriculture would ever have got off the ground. The soils of the Middle East are thin and infertile. When they were ploughed for the first time huge quantities must have blown away on the desert winds. The crops' roots would have fixed them to a certain extent, but without the weeds that rapidly colonise disturbed soil, the bare gaps between the rows would still have been vulnerable to dust-blow, erosion and nutrient loss. Fortunately for the future of soil fertility, most cultivation techniques had an Achilles heel. Late harvest benefited those weeds which produced seeds at the same time as the crop. Cutting with a sickle perpetuated weeds whose seeds were produced at the same height as the wheat ears. Sieving the grains favoured those weed seeds most similar in size to the crop's. This mimicry – a simple expression of the laws of evolution by natural selection – has been the weed's immemorial trick. It can result in extraordinary transmutations. Wild oats have evolved differently shaped varieties to blend in with the crops they accompany. In fields sown with alternate rows of spring and winter barley, the wild oats that sprout amongst the rosettes of over-wintering barley start growing as rosettes themselves, while those amongst the tall spring barley mimic its rapid upward shooting. In the rice paddies of South-East Asia, there are weed grasses so similar to cultivated rice that farmers are unable to distinguish them before the wild grasses bloom. Plant breeders thought they might be able to trick the weed into showing itself by developing a variety of rice with a purple tinge. Within a matter of years, the weed grass had turned purple too. The slight pigmentation that had enabled plant breeders to develop the coloured rice also occurs occasionally in the weed. With each successive harvest it was this strain that was mistaken for rice, and which passed into next year's seed store.

In Mesopotamia there seemed to be no escape from the ingenuity of the thorns and thistles. The more the field-workers attempted to exclude them the more they prospered. Their truculence must have felt like a punishment long before it was detailed as such in Genesis. The values and religious outlook of the Middle East had been transformed by agriculture. The region's original hunter-gatherers had worshipped, or at least respected, animal spirits, independent of humans but biddable. But the first farmers, aware of their new-found powers, needed supernatural beings which would authorise and strengthen their governance of nature. Neither animal spirits nor nature gods could serve this purpose, of course, so the new deities were superhumans, gods made in humans' image, 'shepherds of men'.

But these new powers, and new gods, came at a price. The freedoms of hunter-gathering were replaced by drudgery, division of labour, and, symbolically and literally, by weeds, the tangled baggage of the settled life. For one group of farmers and herders, the tribes of Yahweh – the early Jews – there had been a massive additional trauma. Their heartland in Jerusalem was sacked in 586 BC and the Jews were exiled to the deserts of Babylonia. The Jews interpreted their exile as a punishment, but, with a canny theological twist, saw this punishment as a sign that they had been singled out for divine attention. They rejected the high culture and multiple fertility gods of much of the Middle East, and declared themselves the chosen people of a single God. Monotheism had been invented.

But dissatisfactions with their way of life still rankled, and are worked into the details of the Creation myth they produced. Genesis can be read as an attempt by early Middle-Eastern pastoralists and farmers to explain to themselves why they had to live a life of toil. The conquest of nature – the assumption of knowledge – was both the cause and the form of their punishment. As the twentieth-century philosopher John Passmore argues, the Creation myths are a kind of rationalisation: 'By the time the Genesis stories were composed man had already embarked on the task of transforming nature. [In them he] *justifies* his actions. He did not set about mastering the world – any more than he set about multiplying – because Genesis told him to. Rather, Genesis salved his conscience.'

It's intriguing how many of the central props of the Genesis story – an Arcadian garden and a fall from grace, a serpent, a tree, the eruption of weeds as part punishment, part challenge – recur in the Creation myths of other cultures. It is as if, as symbols, they play some deep structural role in the human psyche. In Classical mythology there is a kind of Eden, too, a pastoral Utopia preserved in a state of perpetual spring, where abundant food crops grew spontaneously, without the need for agricultural toil. Humans are exiled from this paradise also. But here the purpose of the expulsion is not punishment, but *challenge*. The gods believe that putting obstacles in humans' way will encourage them to think and evolve. Weeds are character forming. Virgil's great instructive poem about rural life in pre-Christian Italy, *The Georgics*, describes how Jove willed that 'the path of husbandry should not be smooth', and created a kind of Fall:

> He put foul poison in the serpent's fang
> And ordered wolves to plunder, seas to rage,
> And stripped the leaves of honey, hid the fire,
> And stopped the stream of freely flowing wine,
> So that experience by taking thought
> Might gradually hammer out the arts,

And in the furrows seek the blade of corn.

Ceres taught men to plough with iron, but sowed trouble too, so that it was necessary to 'make unceasing war on weeds':

> . . . and idle thistles reared
> Their bristling heads; the crops began to die,
> A prickly growth of lady's bedstraw sprang
> And caltrop, and amid the shining corn
> Unfruitful darnel and wild oats held sway.

In South America a typical tribal myth about the origin of farming tells of a pre-lapsarian time when humans lived on a diet of fruit and leaves. Then maize was revealed to them by a woman who appeared in the shape of an opossum. The plant was as big as a tree, and grew wild in the forest. But instead of harvesting the seeds, as nut gatherers might, the humans hacked the tree down, and then found that this one-off crop wouldn't satisfy their needs. They were forced to share out the seeds, clear the forest, and plant them out, as the first cultivated crop.

Elsewhere in South America, in the Mato Grosso of Brazil, the anthropologist Claude Lévi-Strauss records an extraordinary myth of the Ofaie-Chavante tribe which is almost the direct converse of Genesis's agricultural story. Honey is classified as a plant in many pre-industrial societies, and in this myth it begins as a cultivated crop, which can be set in the ground to grow and ripen. But it's too accessible and too tempting, and over-consumption soon exhausts the supply. So the animals are directed to go and gather wild honey – 'weed' honey – instead. The drawbacks of cultivation are removed at a stroke. 'There is no doubt about where the originality of this myth lies,' Lévi-Strauss remarks. 'It is, one might say, "anti-neolithic" in outlook, and pleads in favour of an economy based on collecting and gathering, to which it attributes the same virtues of diversity, abundance and preservation claimed by most of the other myths for the reverse outlook, which is a consequence of humanity's adopting the arts of civilisation.'

≈

The weeds of the Fertile Crescent arrived in Britain long before its religions. The first Neolithic settlers from the eastern Mediterranean landed on the south coast around 4500 BCE, a couple of thousand years after the English Channel had opened up. They brought their wheat and barley grains with them in pots or leather bags, and mixed up with them were the seeds of a group of weeds which had never grown in Britain before. In excavations of Neolithic sites dating from around 3500 BCE there is the first evidence of corn poppy, fumitory, charlock and wild radish. By the European Bronze Age (2000 to 500 BCE), they were joined by scarlet pimpernel, black-bindweed, fathen, oxtongue, penny-cress and small nettle.

Just how quickly these arable weeds may have spread from their original field-sites has been suggested by a fascinating experiment in Hampshire. At the Butser Ancient Farm Project archaeologists are experimenting with Bronze-Age farming techniques. They work small fields with facsimiles of contemporary implements and sow ancient crop varieties. In one field of about 900 square yards, cultivated by nothing more than a primitive spade, field penny-cress spread from a single patch about a yard square, and was in all parts of the field within ten years. I have seen the ancient limestone turf of the Burren in County Clare turned into a weed plot after just a few years of regular trampling by cattle. In this extraordinary white-rocked landscape, plants from three climate zones – alpine gentians, Atlantic coastal denizens, orchids from the Mediterranean – have grown together since the end of the last Ice Age. But along tracks trodden and manured by cattle this unique mosaic disappears, replaced by a ribbon of docks and plantains and silverweed, the global signatures of disturbance.

There was not much these early farmers could do about weeds, except pull them out by hand. Or eat them. Some plants – wild carrot, for example – may have suggested themselves as foods by growing very successfully amongst existing crops. Oats – unknown as a crop in the Middle East – was developed in northern Europe from the wild oats which grew as a cornfield weed. Fat-hen was gathered and probably deliberately cultivated for its oily seeds in the Iron Age. It still forms grey-green swarms on manure heaps and well-dunged fields and must have been conspicuous on prehistoric middens. Both the mealy leaves and the farinaceous seeds were made into gruels, and the latter maybe into a kind of unleavened bread. By the early Middle Ages fat-hen – *melde* in Old English – was sufficiently important as a staple to have settlements named after it. The etymology of place names is notoriously difficult territory, but the Swedish historical geographer Eilert Ekwall suggests that Melbourn in Cambridgeshire (*Meldeburna* in Old English) was 'the stream on whose banks *melde* grew', and Milden in Suffolk (*Meldinges c.* 1130) was 'the place of *melde*'. (The modern inhabitants of Milden have no doubts about the origin of their village name. In the 1970s they commissioned a six-foot-tall cast-iron statue of their nominal weed, and erected it on a roadside at the parish boundary.) Elsewhere nettles, chickweed, dock, watercress and mallow survived as subsistence foods long into the days of more organised farming.

But weeds – ubiquitous, obstinate, powerful – had *mana*, and their uses weren't purely domestic and earthbound. In 1950, the perfectly preserved corpse of an Iron Age man was discovered by two peat-diggers in a bog in Tollund Fen in Denmark. The anthropologist P. V. Glob, in his absorbing and lyrical account of the discovery, *The Bog People*, describes how the man's features were so clear and fresh he was at first thought to have died quite recently. 'He lay on his damp bed as though asleep, resting on his side, the head inclined a little forward, arms and legs bent. His face wore a gentle expression – the eyes lightly closed, the lips softly pursed, as if in silent prayer. It was as though the dead man's soul had for a moment returned from another world, through the gate in the western sky.'

But he was 2,000 years old, and round his neck was a tightened noose made of leather thongs. He had been killed by hanging. Equally extraordinary was what was found during the autopsy. Inside the stomach were the remains of the young man's last meal, well enough preserved to be identifiable under the microscope. About twelve to twenty-four hours before his execution Tollund Man had eaten a gruel made of cultivated grains (chiefly barley and linseed) and a huge variety of weeds. Some of these – dock, black bindweed, bristle-grass, corn

WEEDS

chamomile and gold-of-pleasure, for instance – may have been gathered incidentally with the grain. But there was an unusual quantity of knotgrass seeds, sufficient to suggest they'd been gathered deliberately. This was an odd finding. Knotgrass seeds are small and not especially plentiful. The bother of gathering them as a food would hardly be worthwhile. But the weed has an exceptional root system, echoed in its vernacular name 'Devil's lingels', (i.e. the Devil's thongs) – a matted, intransigent network of tendrils that is difficult to disentangle from the soil. Perhaps the knotgrass seeds were seen as a distillation of this insistent occupier of their crop fields, and were gathered as a respectful token.

Two years later another Iron Age man was discovered in a bog at Grauballe, eleven miles east of Tollund. His stomach contents were better preserved and more voluminous than those of his predecessor, and were found to contain no fewer than sixty-three different varieties of seed. As well as the species found in Tollund Man, there was clover, rye-grass, Yorkshire fog, fat-hen, buttercup, lady's-mantle, yarrow and smooth hawksbeard.

What impressed Glob was the absence of any trace of greenstuff or autumn fruits in either of the stomachs. The victims had met their deaths in winter or early spring, before any plants had come into leaf. He conjectures that they died during the midwinter celebrations which were held to hasten the coming of spring, when human sacrifices were often made. This would help to explain the contents of what was quite likely a ritual meal, specially blended from crops and their weed familiars to please the Iron Age goddess of fertility, Nerthus. 'It consisted', Glob argues, 'of an abundance of just those grains and flower

52

seeds which were to be made to germinate, grow and ripen by the goddess's journey through the spring landscape.'

Tollund Man's last supper had a wry footnote a few years later. In the summer of 1954, BBC Television's star archaeologists, Sir Mortimer Wheeler and Dr Glyn Daniel, had a version of the gruel prepared for one of their programmes. They didn't enjoy it, despite washing it down with Danish brandy drunk from a cow horn. The bewhiskered and reliably irreverent Sir Mortimer commented to Daniel that he believed the Bog Man, far from being sacrificed, had committed suicide to escape his wife's cooking.

Weeds and wild plants had lost most of their economic significance as supplementary foods by the Middle Ages. The Celtic fringes clung on to nettle broth and wild garlic, Midlands fieldworkers munched the lemony leaves of sorrel to quench their thirsts, Yorkshire folk made a ritual Lent dish from the spartan leaves of bistort (known as 'Passion Dock' locally), and in times of war and poverty almost anything would get eaten, even down to the hairy strings of goosegrass. But in the peasant economy of Britain at least, bread and cultivated root vegetables took the place of foraged nuts and weed seeds.

What did linger – and continue to prosper even more strongly in the modern era – was an interest in the ritual fascination of foraging, as if ingesting wild plants put you back in touch with your biological roots, with your sense of the seasons, with your whole understanding of food as a product of natural processes. This was always a stronger tradition on the Continent than in Britain. The ancient tradition of *la ceuillette*, the seasonal gathering of wild greens and fungi, still flourishes in south-west France. In the spring, wild leeks and dandelion and the tips of black bryony are amongst the favoured harvests. *La ceuillette* is no longer economically essential, and survives as a rehearsal of ancient rights over the land, a celebration of belonging to one's *pays*. In Crete at Eastertime villagers go out in their Sunday best to gather *stamnagathi*, the bitter rosettes of spiny chicory, in a gesture against the stodginess of winter. In nineteenth-century America, Henry Thoreau celebrated that mysterious quality of 'gatheredness' that clung like a savour to foraged wildings: 'the bitter-sweet of a white-oak acorn which you nibble in a bleak November walk over the tawny earth is more to me than a slice of imported pine-apple'.

A century later another American revived the tradition of transcendental foraging and wrote one of the country's more unlikely best-sellers. Euell Gibbons was born into a poor family, and grew up in New Mexico during the Dust Bowl era. When he was a teenager his father left on a last-ditch hunt for work, and the family of five were left to survive on a handful of beans and a single egg. Euell headed off for the mountains with a knapsack, and came back with it full of edible wild plants. For the next month his family lived wholly on what he provided, and by their own account he saved their lives.

Over the next thirty years Gibbons worked as a cotton picker, a shipyard worker, a beach bum, but always dreamed of being a writer. His attempts at fiction never even made it into print. But on the advice of a literary agent he put together his experiences of wild food foraging in a book with the irresistibly clever title of *Stalking the Wild Asparagus* (1962). It was full of Native American lore, accounts of fruit- and weed-hunting safaris, and extravagant recipes for implausible ingredients (Japanese knotweed jam, burdock-pith preserves). It wasn't about survival, as his early foraging adventures had been, but about reconnecting with the landscape and seasons, and rediscovering the essentials of food in an era corrupted by supermarket culture. It perfectly caught the 1960s middle-class mood of ecological anxiety, and launched still-flourishing foraging cults (Gibbons calls them 'neo-primitive food gatherers') on both sides of the Atlantic. But, as Gibbons recognises in his use of that phrase, the roots of the practice reach very deep, back beyond the Christian demonisation of weeds to the meals fed to the victims of Neolithic fertility rites.

# 4

# *Waybread*

'Mother of worts . . . powerful within'

TO GAZE AT ALBRECHT DÜRER'S extraordinary painting *Large Piece of Turf* (*Das Grosse Rasenstück*, 1503) is to glimpse an imagination piercing through the artistic conventions and cultural assumptions of its time and projecting itself forward three centuries. This is painting's discovery of ecology. This is any corner of any waste patch of land in the early twenty-first century, or at any time. This is a clump of weeds looked at with such reverent attention that they might have been the flowers of Elysium.

The structure of the painting couldn't be simpler. It is the structure of vegetation itself, as if Dürer had stuck a spade at random in the ground and used the slab of turf it lifted as his frame. In the foreground are three rosettes of greater plantain, a weed that has so closely dogged human trackways across the globe that it was also known as Waybread and Traveller's-foot. They're surrounded by wisps of meadow-grass. Two dandelion heads, some way past flowering but still topped with yellow, lean leftwards. At the very rear of the painting – and its only concession to the less than commonplace – a few leaflets of burnet-saxifrage are just visible through the mesh of grass leaves. You observe this community of plants not from above, or any other conventionally privileged viewpoint, but from below. The bottom quarter of the picture is almost entirely devoted to the mottled patch of earth in which the weeds are visibly rooted. The tallest tower above the 'stage' of the picture, as if they were a forest canopy shading their smaller cousins. It is a visually exquisite and scientifically correct composition. What you are looking at is a miniature ecosystem in which every component, from the damp mud at the base to the seeds on the point of flight, is connected.

No one was to take such an intensely grounded view of mundane vegetation again until the early nineteenth century, when the poet John Clare 'dropped down' to marvel at the weeds he loved, and Goethe gave his painter hero Young Werther a transcendental experience while sprawled in the grass: 'I lie in the tall grass and, closer thus to the earth, become conscious of the thousand varieties of little plants . . .'

Dürer's *Turf* is not only the first portrait of a community of weeds, it is the first truly naturalistic flower-painting in Europe, and the herald of a new humanistic attitude towards nature. It had taken more than 300 years from the first stirrings of realism in plant illustration to arrive at this point. In the medieval period illustrations of wild plants appear chiefly in two guises. They are either decorative, ornamenting the borders of Books of Hours and courtly pictures of flowery meads, or they are functional, visual pegs for descriptive accounts of medicinal species. In neither context is there any pretence at botanical accuracy. This scarcely mattered when a vague aura of seasonal floweriness was all that was wanted. But the treatment of illness was another matter. Plants probably made up more than 90 per cent of all the substances presumed to have healing properties. Many were prescribed according to bizarre amalgams of magic and pre-scientific beliefs about the working of the human body, especially the theories of the four humours put forward by ancient Greek and Roman physicians such as Hippocrates and Galen. But every system of treatment, however wrong-headed, depended on the identification of the 'right' plant. This was the business of the herbal, a text which gave descriptions of medicinal plants and how they were to be used for particular disorders.

Yet until the sixteenth century the illustrations in herbals were often stylised to the point of abstraction. It wasn't that medieval illustrators lacked the technical skills: contemporary sketches of human and animal figures are often lively and imaginative without sacrificing any of their persuasive realism. But their plants have the look of motifs from a pattern book. They are simplified and symmetrical. Flowers are amorphous blobs on the end of stiff stalks. Roots are variations on the single theme of carrot. It is as if plants lacked some essential animating spirit with which the artist could empathise.

But it is also the case that the illustrators often had no idea what they were meant to be illustrating. The tradition of acute and thoughtful observation of nature begun by Aristotle and Theophrastus faded with the collapse of the Greek and Roman empires. In medieval Britain especially, Anglo-Saxon magical beliefs and an intensely authoritarian Christian Church actively discouraged objective enquiry into the lives and properties of plants. It was as if questioning the workings of nature, rather than simply accepting conventional priestly dictums, was a kind of blasphemy, a challenge to God's ordering of the world.

Curiously, the teachings of the pagan Classical authors were highly valued, at least when it came to identifying medicinal plants. They were seen as possessing wisdom long lost in Dark Ages Britain. Much of what passed for intellectual effort in medieval botany and medicine was spent in attempting to understand and reinterpret Classical texts. In practice this meant endless copying and recopying, with all the potential for errors this allowed. This was a job largely done by the monastic orders. The monks could read Latin, often knew a little medicine, and would in all probability have a herb garden, used for treating their own and the surrounding community's illnesses.

Their most important source of information on healing plants was *De Materia Medica*, written in Greek in the first century AD. Every European herbal for the next 1,500 years was in some way inspired by or derived from this single hallowed source. The author of *De Materia Medica*, Pedianos Dioskurides, known today as Dioscorides, was probably an army doctor from Asia Minor, and an accomplished botanist. His text stresses the fundamental importance of getting to know plants in the field:

> Now it behoves anyone who desires to be a skilful herbalist to be present when the plants first shoot out of the earth, when they are full grown, and when they begin to fade. For he who is only present at the budding of the herb, cannot know it when it is full-grown, nor can he who hath examined a full-grown herb, recognise it when it has only just appeared above ground. Owing to changes in the shape of leaves and the size of stalks, and of the

> flowers and fruit, and of certain other known char-
> acteristics, a great mistake has been made by some
> who have not paid proper attention to them in this
> manner.

Alas, later writers and editors paid little attention to his fastidious instructions. There is a superb edition, the *Codex Vindobonensis* from sixth-century Constantinople, which has some naturalistic portraits amongst its 400 full-page coloured illustrations, but many of the early editions were not illustrated at all. Others contained stylised drawings of notional plants or crude copyings from earlier illustrators or plain botanical fantasies. This process continued down the centuries, with the copies becoming increasingly unrecognisable as real plants. Few illustrators seemed inclined to go out and draw from life, partly because they were not always able to identify what species the Classical writers were talking about. But there is also a sense that this was unnecessary or inappropriate, that the checking of Classical edicts against the commonplace realities of the world was a kind of impertinence.

In the British herbal tradition, the first glimpse of a different approach came early in the twelfth century. In about 1120 the monks of Bury St Edmunds Abbey in Suffolk produced a version of the *Herbarium* of Apuleius Platonicus, the earliest known copies of which date from the fifth or sixth century. The Latin text is an unoriginal compilation of medicinal prescriptions from Dioscorides and other Greek sources, but scattered amongst them is a handful of plant illustrations with a freshness and realism that had not been seen before in northern Europe.

The manuscript of Apuleius' *Herbarium* survived, and

is now in the Bodleian Library at Oxford. It's a curious and unassuming book, not much larger than a modern paperback, and covering little more than a hundred species on its parchment pages. The text, where you can make sense of the imperfect Latin, reads at times like a book of spells. For mugwort it suggests that 'if a root of this wort be hung over the door of any house then may not any man damage the house'. For castor-oil plant: 'If thou hangest some seed of it in thine house or have its seed in any place whatsoever, it turneth away the tempestuousness of hail, and if thou hangest its seed on a ship, to that degree wonderful it is, that it smootheth every tempest.' Many of the illustrations are fantastical, too, a conventional mixture of crude copy and stylised invention. The hard-to-guess 'Herba Lapan' (names are given in Latin, Gallic, even Egyptian) is painted with alternate green and bright blue leaves, their veins tricked out with gold leaf. Asphodel is shown upside down, with a root like a dragon. 'Herba Gram' (probably a legume) wanders thinly about like a line in a Paul Klee sketch.

But there are several different hands behind the illustrations, and maybe a dozen of them were clearly sketched from life, probably by inquisitive monks who went out into the Abbey gardens and the Suffolk countryside and looked at growing plants with an attentiveness that would have gratified Dioscorides. It's hardly surprising that almost all these naturalistic illustrations are of common, accessible and easily identifiable weeds. Plantain is there, vervain, mint, and a dandelion (recommended for urinary problems) with a splendidly splayed rosette of leaves. Chamomile flowers have perfect yellow tea-cosy centres, and the trefoil leaves and compact flower tufts of red clover are unmistakable. Some of the drawing is quite skilled technically, as, for example, that of the complex labiate flowers of hedge woundwort (recommended for cuts and ulcers). But the outstanding piece is the bramble, which winds its prickly way around the text's dark prognostications on the bites of a *serpens*. The clusters of blackberries still look munchable, almost dewy, 900 years on. There are, as always in the countryside, a few unripe red fruits in each bunch, and the gloss on the blacker berries is captured by a single spot of blue-grey paint in the middle of each drupe.

The gap between the Bury St Edmunds herbal's plantain and dandelion, crudely but fondly sketched and written about in language not much removed from magical incantation, and Dürer's evocative and ecologically informed drawing of the same species growing together in a patch of turf, is the space where the medieval attitude towards weeds took shape. Weeds, quite simply, were present in every aspect of life, and were still regarded as part of Adam's curse. They were of unknown origin, and believed to be capable of magical transformations. The same plant could poison you or cure your pain. A food crop could 'degenerate' and metamorphose into a choking vegetable plague. There was, to the medieval mind, no point in looking for rhyme or reason in these perverse, punishing plants. The mysterious irrationality of the physical world was humanity's bitter legacy from the Fall, and had to be endured.

The local historian, Revd Foster Barham Zincke (chaplain to Queen Victoria), found this stoical attitude persisting in rural Suffolk into the late nineteenth century:

I have heard it confidently announced as if there could be no doubt about it, that weeds are natural to the ground, in the sense that the ground originates them; and that no man ever did, because no man ever could, eradicate them. They spring eternal from the ground itself, not at all necessarily from the seeds of parent weeds . . . To this ignorance is super-added in the case of weeds a theological conception, that the ground has been cursed with weeds as a punishment for man's disobedience. It has therefore ever borne, and will ever continue to bear, for the punishment of the husbandman (but why should the husbandman only be punished?) thistles and poppies and speargrass.

Thomas Tusser's account of agriculture's struggle with the travails of the world, *Five Hundred Points of Good Husbandry*, was written at the tail-end of the medieval period, and is free of the fearful resignation of that earlier time. Instead it bounces with the confidence and practicality of the dawning age of science. Its air of self-assurance is boosted by Tusser's decision to cast the whole text in jaunty rhyming couplets. Nevertheless it's one of the only first-hand accounts to give a clear picture of the medieval farmer's round, and of how he regarded weeds. As unrelenting trouble, Tusser seems to say, but no real match for a canny Essex yeoman. Early summer was the time to attack them, especially after a shower had loosened their roots:

In May get a weed-hook, a crotch and a glove,

And weed out such weeds, as the corn do not
    love.
For weeding of winter corn, now it is best;
But June is the better for weeding the rest.

The May-weed doth burn, and the thistle doth
    fret;
The fitches pull downward both rye and the
    wheat:
The brake and the cockel, be noisome too much;
Yet like unto boodle, no weed is there such.

Tusser's weed-hook was a pair of long-handled prongs, one of them forked at the end, the other tipped with a metal hook, with which each weed was individually twisted out. (Hoeing – though it was a technique familiar to Native Americans and had been practised in Europe since at least the time of Virgil as a way of aerating the soil – wouldn't be adopted for weed control in Britain for another two centuries.) The social historian Dorothy Hartley followed the laborious progress of the weeder in her book *The Land of England*. She doesn't explain how she built up her exact, footstep-by-footstep account, but it was likely pieced together from clues in old paintings and intuitive reconstruction from the vestiges of ancient practices that she glimpsed during a lifetime's fieldwork. This is what she wrote:

He uses two sticks: with the first, hooked stick he plucks the weeds out from under the corn stalks, and with the second, forked stick pins the weed's head down under the fork. The weeder then steps one pace forward, placing his foot on the head of the weed, and, with this forward movement, swings the hooked stick round behind him, lifting the root of the weed high out of the ground, before dropping it in line. In this way each pulled-up weed is shaken clear of the soil, and laid with its root over the buried head of the previous weed. Thus, as the weeder goes along the line of the furrows he lays a mulch of decaying weeds alongside the roots of the corn, and forms a line between the rows at least as wide as his foot. Weeding employed a definite rhythm, and the feet of the weeder formed the lines on which much of the reaper's work depended.

It sounds like a recipe for customised control, every weed being extracted as precisely as a bad tooth. But medieval weeding was no match for the evolutionary wiles of the most aggressive species. In removing visible weeds in a particular season, it gave a selective advantage to varieties with, for example, deep or spreading roots, which the hook failed to extract; to varieties with the ability to flower and set seed before the weeder came, and which in effect used him as a planter when he laid out the pulled plants and scattered their seeds on the soil; to varieties that, conversely, sprang much later in the year, maybe between harvest and first ploughing. The goosegrass that grows in cultivated fields, for example, is quite different from that in hedges. It germinates at different times, has seeds closer in size to crop seeds and grows with a more creeping habit – adaptations to weed-control methods that go back two or three thousand years.

Persistent hand-weeding does eventually weaken most species, especially annuals that can be extracted or killed before they set seed. Deep-rooted and expansive perennials are less affected, and are often inadvertently spread about. Many of these species – such as creeping buttercup and silverweed – are able to regenerate from tiny fragments of root or runner, so a weed-control technique which was based, in essence, on pulling a weed free of most (but never all) of its roots and then dumping it back on the soil, simply multiplied the number of potential plants. The sorrows of farming, as Genesis promised, would continue 'all the days of thy life'.

Stinging nettle is a perennial species that increased enormously as a consequence of cultivation and weeding. Its natural habitat is fertile, muddy, slightly disturbed ground, especially amongst the lush herbage of nutrient-rich silt in river valleys and woodland clearings manured by grazing animals. It adapted very easily to the enriched soils of arable and pasture land (especially as silt from ponds and rivers was frequently spread on fields to boost fertility), and to man-made sites rich in nitrogen and phosphorus – middens, bonfire sites, churchyards. Nettle spreads by seed, and by the aggressive growth of its underground stems, which can advance more than two feet a year. Even severed fragments of these stems can spread horizontally, send down tough, fibrous roots, and eventually break through the surface to form new leaf-bearing stems. Vast colonies can build up from this radiating underground structure. Phosphates in the soil persist for exceptionally long periods, and the wooded sites of Romano-British settlements on the Grovely Ridge near Salisbury are still dense with nettles thriving on the remains of a human occupation that ended 1,600 years ago. (They are flourishing in Wiltshire's modern landscape, too. Fertiliser run-off from the county's huge expanses of industrial arable farming, plus phosphates from household detergent, drains off into the River Kennet. In summer a twelve-mile stretch is an almost continuous double ribbon of nettles, some of them ten feet high.)

The bindweeds are a family that have perfected an extensive and daunting battery of survival techniques. Their sinuous roots and climbing stems, which can smother other vegetation, earned them the uncompromising vernacular name 'Devil's guts'. Before the development of chemical herbicides, the field species, *Convolvulus arvensis*, was amongst the most intractable of arable weeds. It's a beguilingly attractive plant, with pink, white or candy-striped bellflowers which have a light almond fragrance in the sun, and whose nectar attracts a large number of insect species. The twining stems may be a clue to its wild origins. They can reach at the most about three feet in height, nothing like the tree-climbing hawsers of the hedge bindweed, and suggest that Devil's guts may have begun its conquest of farm and garden from areas of disturbed soil studded with low bushes – the kind, for example, that develop at the foot of unstable cliffs. Field bindweed's most natural-looking modern habitat is short, stony grassland close to the sea. But the plant we know today is so exquisitely engineered to cope with the pressures of cultivation that it may have continued to evolve through the past few thousand years of the agricultural era.

Field bindweed has an almost foolproof insurance portfolio, a range of reproduction and regeneration techniques to meet every possible contingency. Each plant produces about 600 seeds, some of which germinate in the summer and some in autumn. Or, if buried deeply enough, at any time over the succeeding forty years. Once the seedling is established and rooted it extends horizontally by means of underground stems. The whole underground system may spread over thirty square yards in a single season, and the vertical roots penetrate downwards more than eighteen feet. New above-ground shoots can spring either from the underground stems or directly from the roots. Cutting the roots with a hoe or plough temporarily weakens the plant, but also promotes new shoots. The response by the plant is fast and decisive. Within a few seconds a milky latex oozes out of the wound, and clots over the cut surface to form an antiseptic seal, a callus. Within days dormant shoot-buds close to the wound have begun to swell and form new roots and leaf stems. This happens with even the tiniest fragment of any part of the plant. A bindweed root or stem chopped into a hundred pieces by a frustrated gardener is simply the starting point for a hundred new plants.

Above ground, the tip of the twining stem hunts for light, curling round any vertical objects – other plants included – for support. (In the laboratory, bindweed shoots can find their way to a light source through a maze of blackened tubes.) Any damage to these host plants is collateral: the bindweed just wants scaffolding. If the twining stems are partially buried by soil or stones, they can take root. If they're repeatedly cut off the plants compensate by taking on a bushy form and generating multiple branches. If they're eaten by cattle, chemicals in the stem recognise the growth hormones in the animal's saliva, and are stimulated into even faster regrowth.

There is, surprisingly, a hopeful lesson in this story of bindweed's effortless shape-shifting. For all its formidable survival strategies it doesn't prosper beyond cultivated or disturbed ground. You won't find it in woodland because it is absolutely dependent on light. It can't penetrate the settled plant communities of old pastures and meadows. It does sometimes invade new lawns, but a year or two of constant mowing will starve even its cussedly vigorous root system. Devil's guts is a supreme survivor, but it is not a superweed.

~

Despite the seemingly diabolic powers of weeds like nettle and bindweed, none, so far as I know, was ever tried for witchcraft or blasphemy or outlandish behaviour. They were lucky to escape. In the medieval period – until the mid-nineteenth century, in fact – any living thing might be taken to court if it was thought to be violating God's laws or society's codes. In 1499 some sparrows were excommunicated for depositing droppings on the pews in St Vincent, in France. In 1546 a band of weevils were tried for damaging church vineyards in St Julien. Such trials were rife in the sixteenth century, and the distinguished French lawyer Bartholomew Chassenée rose to fame as an advocate for animals. His work is commemorated in Julian Barnes's mischievous short story 'The Wars of Religion', in which excommunication is sought for a colony of woodworm which had gnawed away the supporting legs of the Bishop of Besançon's throne, causing him to be 'hurled against his will into a state of imbecility'. Chassenée negotiates a gentler outcome for the insects. They are excused punishment, provided the local inhabitants 'set aside for the said *bestioles* alternative pasture, where they may graze peacefully without future harm to the church of St Michel, and that the *bestioles* be commanded by the court, which has all such powers, to move to the said pasture'. (The precocious idea of a reservation for undesirables prefigures the modern notion of weed 'corners' in gardens and unploughed strips beside arable fields.)

The worst that pious medievals did to weeds was to shout bad names at them. There are at least twenty species with vernacular tags (now mostly obsolete) that identify them as plants of the Devil. Mayweed was Devil's daisy. Corn buttercup was Devil's claws, Devil-on-all-sides, Devil's coachwheel and Devil's currycomb (mostly references to the shape of the seeds). Deadly nightshade was Devil's rhubarb and Devil's berries. Mullein – Devil's blanket (from the downy leaves). Ground-ivy – Devil's candlestick. Dodder – Devil's thread and Devil's net (also Hellweed and Hellbind). Bird's-foot trefoil – Devil's fingers (but also Our Lady's fingers). Greater stitchwort – Devil's corn and Devil's skirt buttons. Shepherd's needle – Devil's needle. Henbane – Devil's eye. Nettle – Devil's leaf. Cow parsley – Devil's meat and Devil's oatmeal. Dandelion – Devil's milk-pail (from its white latex). Redshank – Devil's pinch. Black-bindweed – Devil's tether. Corn poppy – Devil's tongue. Fool's parsley – Devil's wand. And sun spurge, a modest annual all of nine inches tall, was given the most extravagant accolade of all, being 'the Devil's apple tree' in parts of Scotland.

But whether any of these weeds were genuinely regarded as the Prince of Darkness's familiars is doubtful. The satanic epithets were probably just terms of respectful joshing, as in 'that little devil'. The supernatural associations of plants, as we shall see, lay in a quite different quarter.

A more matter-of fact litany of names (it amounts to the first blacklist of weeds) is given in John Fitzherbert's *Complete Boke of Husbandry* in 1523. 'In the later ende of May is the tyme to wede thy corne,' he wrote, prefiguring Tusser. 'There be diverse manner of wedes as Thistles, Kedlokes [charlock], Dockes, Cockle, Darnolde, Gouldes, and Dog fennel.' This last was Tusser's 'May-weed' (stinking may-weed today), whose acrid secretions could blister the skins of fieldworkers. Goulde was the yellow-flowered corn marigold, Tusser's 'Boodle', also known as guildweed, or gold (gold was boodle long before the days of cartoon burglars) and one of the most intractable of medieval weeds. It was regarded as such a problem in the twelfth century that Henry II issued an ordinance against it, an enactment whose restrictive powers were not rivalled until the Weeds Acts of the twentieth century.

But the most serious intruder, in terms of its impact on human beings, was cockle. The name referred to two quite different species, linked by the fact that their seeds, milled with the wheat, made flour foul tasting and often toxic. Corncockle, a member of the pink family Caryophyllaceae with exquisite purple flowers that open from their buds like unfolding flags, has seeds which ripen at the same time as wheat, are of much the same size and weight as wheat-ears, and which aren't easily separated out by winnowing. They turn flour – and the bread made from it – grey. The noxious glycosides in the plant, known as saponins, enter the bloodstream and cause a breakdown of red blood corpuscles and other cells. The condition (still common in India) is known as githagism, after the plant's Latin name *Agrostemma githago*, and is characterised by lassitude, yawning, loss of weight and enteritis.

The other cockle – darnel-cockle, Fitzherbert's 'Dar-nolde' and the Bible's tares – was the grass now known simply as darnel, a close relative of rye-grass. When its seeds found their way into bread, it caused another raft of symptoms in the hapless eaters – ringing in the ears, nausea, impaired vision, abdominal pains and diarrhoea, though these were rarely long-lasting.

The remarkable thing is that just after the weedings of late May and early June many of the species ripped out by the weeding hook and buried unceremoniously in the soil were resurrected – or had their benign essences liberated – in a Midsummer fertility ritual. On Midsummer's Eve, great bonfires were lit in the countryside, and bundles of wild herbs thrown on them. Most of the plants were agri-cultural weeds, including St John's-wort, corn marigold, corn poppy, mayweed, mugwort, ragwort, plantain and vervain.

Midsummer is one of the great hinges of the natural year in northern Europe. It's the time of the solstice, when the sun seems to 'stand still' before its long decline into winter and the night is so brief it seems to merge with the day. For the medievals it was a moment when the barriers between humans and nature – and between the worka-day world and the domain of magic – might dissolve. No wonder it was a time of elaborate pagan ritual – even in notionally Christian Britain – well into the nineteenth cen-tury (and into the twenty-first in a few places). The future might be glimpsed, fertility guaranteed.

The most important principle behind magical beliefs in almost all pre-scientific cultures is sympathy. Put sim-ply, like would cure – or provoke, or resonate with, or sometimes repel – like. Passing a child with a broken limb through the split trunk of a tree, and then binding up the trunk, would speed the healing of the bone. The mating rituals of animals, if mimicked by human dancers, would make the animals more fertile, and maybe the humans too. A mirror hung at the garden gate would reflect, and thus deflect, thunderclouds. At the core of sympathetic magic was the idea of analogy, the belief that all the different components of the universe were connected – not physically or ecologically, but through 'influences' suggested by their superficial appearances.

So a blazing fire would bolster the heating power of the sun. The Midsummer fire festivals, which were held throughout Europe, fulminated with sympathetic magic. The flamboyant displays of light and heat would encourage the sun at the precise moment its own strength began to wane. As a bonus they would cauterise pestilence and 'noisome influences' in the earth. The fires still survive in some regions. In the Pyrenees they are lit by parish leaders, or sometimes by the inhabitants of a single street. In Sweden the huge bonfires are accompanied by maypoles, which, as in some other northern countries, appear in late June rather than May. In Cornwall the tradition has recently been revived, and a chain of fires can be seen crowning the Cornish hills at Midsummer, alongside the processions of flaming tar-barrels carried aloft on poles.

In parts of Britain the fires were lit on the windward sides of fields, so that the purifying smoke would blow over crops and cattle. The choice of which plants to burn reflected the solar magic of the fire itself. They were mostly summer-flowering species whose white, red or yellow blooms mirrored the shape and colour of the sun. Perhaps their origins, as familiars of the crops, was also significant, as in the use of knotgrass in Neolithic fertility rites as practised on Tollund Man.

During the Middle Ages, the Midsummer fires were appropriated by the Christian Church, and said to be lit in honour of St John, whose festival (celebrating his birthday, not his death) is Midsummer Day, 24 June. But the pagan content of the rituals didn't change. Young girls still divvied their future lovers, and virile farm-boys leaped through the flames. And the distinction of receiving the saint's name fell on the most magical of all the plants in the fire-mix. St John's wort (its pre-Christian names are lost) is the very type of a sun-herb, a plant of high summer whose flower is a bright yellow star. It also has an extra feature which, to the medieval mind, marked it out as an obvious supplication to the life-light of the sun. Each leaf is covered in tiny transparent dots (the perforations of the Latin name, *Hypericum perforatum*) and held up against the sky, the sun's rays prick through, like dapple in a springtime wood.

If the medieval herbals, with their muddled translations of the Classics and phantasmagoric view of the natural world, expressed the contemporary Christian view of the workings and influence of plants, rituals such as the Midsummer fires are probably the best insight into the ordinary person's beliefs. There is also a small group of Anglo-Saxon herb books which give clues to popular views about the power of plants, eclectic concoctions of Celtic lore and sympathetic magic and backwoods Christian ritual. In this period plants were no longer seen as being inhabited by gods. But they did possess powerful essences which could deflect, or protect against, malign influences. The air was full of such dangerous forces, which could provoke anything from a failed harvest to a husband's adultery. Disease was often attributed to a mysterious and amorphous substance called 'elf-shot', which was targeted at hapless mortals by supernatural beings. A remedy for elf-shot in *The Leech Book of Bald*, compiled roughly two centuries before the Bury St Edmunds *Herbarium*, is an extraordinary confection of plant witchcraft and Christian rite:

> Go on Thursday evening when the sun is set where thou knowest that helenium [probably any daisy family member – especially disc-flowered sun-weeds like chamomile] stands, then sing the Benedicite or Pater Noster and a litany and stick thy knife into the wort . . . go again as quickly as thou art able to church and let it lie under the altar with the knife; let it lie till the sun be up, wash it afterwards, and make into a drink with bishopwort and lichen off a crucifix; boil in milk thrice, thrice pour holy water and sing over it the Pater Noster, the Credo and the Gloria in Excelsis Deo, and sing upon it a litany and score with a sword round about it on three sides a cross, and then after that let the man drink the wort; Soon it will be well with him.

Closely allied to the idea of elf-shot was 'flying venom', a stream of dangerous ectoplasm which was blown about in the wind. Sometimes this was thought to have originated in fragments of the serpent or worm which the god Woden had cut into nine fragments – a doctrine which takes the idea of disease causation back to the earliest metaphors for evil. The most potent protection was to employ a charm or potion based on the Anglo-Saxons' nine sacred herbs, which included several familiar weeds: mugwort, plantain, stime (watercress), maythen (mayweed or chamomile) atterlothe (probably betony) wergulu (stinging nettle), chervil, fennel and crab apple. The fact that weeds might be simultaneously a curse and a benediction wasn't a cause of confusion. As today, it was a matter of context. In the soil, they were trouble; in the sickroom, a cure. Their ubiquitousness and obstinate power in the fields may even have strengthened their healing image.

Plantain, 'the mother of worts', is present in almost all the early prescriptions of magical herbs, back as far as the earliest Celtic fire ceremonies. It isn't clear why such a drab plant – a plain rosette of grey-green leaves topped by a flower spike like a rat's-tail – should have had pre-eminent status. But its weediness, in the sense of its willingness to tolerate human company, may have had a lot to do with it. The Anglo-Saxon names 'Waybroad' or 'Waybread' simply mean 'a broad-leaved herb which grows by the wayside'. This is plantain's defining habit and habitat. It thrives on roadways, field-paths, church steps. In the most literal sense it dogs human footsteps. Its tough, elastic leaves, growing flush with the ground, are resilient to treading. You can walk on them, scuff them, even drive over them, and they go on living. They seem to actively prosper from stamping, as more delicate plants around them are crushed. The principles of sympathetic magic, therefore, indicated that plantain would be effective for crushing and tearing injuries. (And so it is, to a certain extent. The leaves contain a high proportion of tannins, which help to close wounds and halt bleeding.)

The elastic powers of plantain extended beyond first aid, though. It was also a divination herb, stretching sight into the future, and was used especially at that time when the membrane between the human and supernatural worlds was at its thinnest. On Midsummer Eve in Berwickshire the flowering stems were employed by young women in a charm which would predict whether they would fall in love. It was a delicate, almost erotic process in which the sexual organs of the plantain were used as symbolic indicators. Two of the 'rat's-tail' flowering spikes were picked, and any visible purple anthers (the pollen-bearing tips of male organs) removed. The two spikes were wrapped in a dock leaf and placed under a stone. If, by the next day, more anthers had risen erect from the flowering spikes, love was imminent.

Even in the seventeenth century London plantain divination was being practised at Midsummer. John Aubrey was walking in the pastures by Montagu House on 24 June 1694 when he saw some two dozen women, 'most of them well habited, on their knees very busy, as if they had been weeding'. He asked a young man what they were up to. He was told that 'they were looking for coal under the root of a plantain, to put under their head that night, and they should dream who would be their husbands'. The Anglo-Saxons' 'Lay of the Nine Herbs' had spelt out all these resilient, apotropaic, future-foretelling powers nearly a thousand years before:

> And you, Waybread, mother of worts,
> Open from the eastward, powerful within,
> Over you chariots rolled, over you queens rode,
> Over you brides cried, over you bulls belled;
> And these you withstood, and these you
>     confounded,

So withstand now the venom that flies through
   the air,
And the loathed thing which through the land
   roves.

~

A thousand years on we know that a kind of flying venom
does exist – not as something which weeds can help repel,
but which they emanate themselves. The air and the soil
are busy with constant streams of chemical messages –
plant pheromones – designed to deter predatory insects,
seduce pollinators, kill off competitors, encourage companion plants and warn other plants of insect attack. The
pheromones may be volatile, and transmitted from leaves
through the air, or water-soluble root-exudations that
leach through the soil. The more plants that are involved,
the more complex the play of messages becomes, and
in long-established plant communities this chemical
polyphony may be one of the devices by which interlop-
ers like weeds are kept out. But in disturbed ground with
few established plants, there is little existing activity, and
weeds can begin their own chemical barrage to suppress
competitors. Field bindweed and creeping thistle exude
pheromones which inhibit the germination of most grain
crops. In an elegant series of experiments the roots of the
American weed quackgrass have been shown to suppress
the growth of maize not just conventionally, by monopo-
lising nutrients in the soil, but by actively poisoning the
corn. Every part of the quackgrass plant produces this
toxin, so some of the effects may be airborne. Even the
seeds of some weeds – for instance fat-hen, thornapple

and cockspur grass – leach chemicals into the soil, and can suppress the germination of crops such as cabbages, carrot and tomato. But the chemical traffic isn't one way. Wheat, oats and peas suppress fat-hen. Exudations from cotton plants stimulate the germination of the parasitic striga (sometimes known as witchweed) even though cotton is not one of its hosts.

The invisible chemical transactions between weeds and other plants are still barely mapped, but 500 years ago, herbalists had already intuited something of this chemical messaging with another troublesome weed. The dodders are a remarkable group of plants, not only entirely parasitic but existing pretty much without roots. They prey on thyme, legumes and, in the Middle Ages especially, on flax. Flax-dodder could wipe out an entire crop.

Dodder's growth is mysterious and unearthly. When its seeds germinate in late spring, they produce fine red or yellow threads. The threads have no leaves, no chlorophyll and no attachment to the ground. They seem to slither more than grow. The tip of the seedling is raised above the ground, and edges forward. Meanwhile the whole thread is spiralling, so that in fast-frame it seems to be advancing with the motions of a snake. When the dodder meets and recognises a suitable host it begins to twine round it, alternating a series of loose coils with a series of tight ones. From these emerge rows of tiny suckering spikes, which penetrate the tissues of the host plant and provide a conduit for the extraction of water and nutrients. When dodder attaches itself to large woody plants like furze or heather, it weakens the host but rarely kills it. But with smaller, fleshier species, such as flax, it can often prove fatal – though not, of course, before it has seeded itself and ensured its offspring a future.

The first man to look inquisitively at dodder was William Turner, a Cambridge-educated naturalist, cleric and physician, described by the distinguished British botanist John Gilmour as 'the first in Britain to light his torch at the flame of the pioneer herbalists on the Continent, and, breaking away from authority and superstition, to describe British plants from his own observation and experience'. His herbal was published in 1543, just thirty years after Dürer's illustrative breakthrough. In it he notes dodder's dependence on other plants: 'Doder growth out of herbes and small bushes as miscelto growtheth out of trees. Doder is lyke a great red harpe strynge and it windeth about herbes foldynge mych about them . . . The herbes that I have marked doder to growe most in are flax and tares.' John Gerard, in his 1597 herbal, went further and suggested that 'the nature of this herb changeth and altereth according to the nature and qualitie of the herbs whereupon it groweth'. Modern botanists have found that dodder does indeed exist in a number of races and types, and that each one is adapted to recognising the chemical signature of its particular host. The dodder appears to 'smell' its host, picking up the unique cocktail of volatile chemicals released by the leaves. At Pennsylvania State University, biologist Consuelo de Moraes has studied the tracking techniques of the variety of dodder that preys on tomatoes (dodder is known as strangleweed and lovevine in the USA). She found that the growing tip of the plant rotates slowly, 'sniffing' for the host, and then aims decisively for it. It ignored ersatz tomatoes fashioned from red felt and pipe cleaners, and globes of coloured liquid, so was not reacting to colour cues. But when de Moraes extracted scent chemicals from actual tomatoes and applied them to a piece of rubber, the dodder immediately shot out tendrils in its direction.

~

There is an ironic footnote to this story of the shifts between the superstition of the Dark Ages and the open-minded curiosity of the Age of Reason. It is, in a literal sense, a cloistered story. Both monastery and university were physically characterised by walls. Walls enclosed their herb gardens and, symbolically, enclosed their knowledge, made it clear to the outside world where intellectual authority lay. Plants with the potential to become weeds are, of course, defined by their contempt for boundaries. What happened in monastic herb gardens was that some of the medicinal herbs took to growing *in* the walls. They used them as stepping-stones into the outside world, and into a more popular consciousness.

The monasteries were themselves institutions for traffic in weeds. Some, such as the Cluniac Order (a branch of the Benedictines), had abbeys in the far south of France, and would have brought medicinal herbs over with them. Amongst these were species that, like the weeds of agriculture, were natives of dry rocky places in the Mediterranean, and they found the dry stonework of the walls designed to keep them in was a congenial springboard for finding their way out.

Remnants of the walls of Bury St Edmunds Abbey, where the Apuleius Platonicus *Herbarium* of 1120 was produced, still survive, and are adorned by weeds. Some are comparatively new (buddleia, for instance), showing that walls are also devices by which plants can get in. But some – feverfew (for headaches), greater celandine (for eye complaints) and wallflower (for ulcers and for help-ing 'the chops or rifts of the fundament') – might just be descendants of plants that grew in the Abbey physic gar-den 900 years ago. All are now common on stonework and in waste places throughout Britain.

If the cultivated field provided a portal for weeds in the countryside, the boundary wall and the road were their gateways to the urban world. Whatever institutions humans created to preserve their civilisation from the wild, weeds found ways of exploiting them.

# 5

## *Self-heal*

### The medicinal weed

ONE WEED SPECIES whose reputation prospered during the age of magic was mandrake. This legendary member of the nightshade family is a widespread weed in the Mediterranean region, cropping up in olive groves and fallow land and wheatfields that aren't deeply ploughed. It is mysteriously attractive. A rosette of large, wrinkled, dark green leaves lies flat on the ground, and in the centre, often in late autumn, a cluster of deep purple flowers appears. But it was the roots that fascinated herbalists. They are deeply forked and skin-coloured, and occasionally resemble a crude human figure, a manikin, complete with genitals. So, by sympathetic magic, they were regarded as an aphrodisiac, a cure for sterility, even as the source of a potion for driving out demons.

A thicket of mythology surrounded the mandrake's life and uses. It was said to grow best under a gibbet. If the decaying body that leached compost onto the plant was a man, the mandrake root would grow male genitals; if female, the root would take on the shape and detail of a woman. And since the root was an analogue of a human being, special distancing techniques were advised for picking it, lest the gatherers be implicated in – and magically avenged for – a kind of murder. The plant would shriek when it was pulled up, and pickers were advised to tie a dog to the stem and let it do the job. And die, if necessary.

As a matter of medical fact the mandrake does contain narcotic alkaloids. In Classical times it was used as a mild anaesthetic during surgery. In England the herbalists who peddled it could not always get hold of the authentic root, and substituted roots of other plants, often of the far more toxic white bryony. In fact it looks as if many of the superstitions associated with picking the roots – the fatal shriek, the haunting of execution sites – were propagated by professional herb gatherers and merchants, to deter outsiders from poaching their lucrative crop.

The clear-sighted William Turner was scathing in his denunciation of mandrake quackery:

> The rootes which are counterfited and made like little puppettes and mammettes which come to be sold in England in boxes with hair and such forme as man hath, are nothyng elles but folishe fened trifles and not naturall. For they are so trymmed of crafty theves to mocke the poore people withall and to rob them both of theyr wit and theyr money. I have in my tyme at diverse tymes taken up the rootes of mandrake out of the grounde but I never saw any such thing upon them or in them as are in and upon the pedlers rootes that are comenly to be solde in boxes.

The problem was that the dawning interest in the observation of plants, typified by Turner, was seized on by the herb peddlers – and to a certain extent by the Church – and used to underpin an early prototype of the theory of

Intelligent Design. The Doctrine of Signatures was sympathetic magic tidied up, stripped of its blatant magical influences, and given Christian authority. What it taught was that God had 'signed' plants, with certain suggestive shapes and colours, say, in order that humans could 'read' the illnesses they were designed to ease. So looking closely at plants to interpret their design could now be seen as an act of Christian devotion.

The Doctrine's most evangelical proponent was the seventeenth-century Oxford-educated herbalist William Coles. He outlined the system in *Adam in Eden, or Nature's Paradise* (1657).

> Though Sin and Sathan have plunged mankinde into an Ocean of Infirmities, Yet the mercy of God which is over all his Workes Maketh Grasse to grow upon the Mountaines and Herbs for the use of Men and hath not onely stemped upon them (as upon every man) a distinct forme, but also given them particular signatures, whereby a Man may read even in legible Characters the Use of them. Heart Trefoyle [spotted medick] is so called not onely because the Leafe is Triangular like the Heart of a Man, but also because each leafe contains the perfect Icon of an Heart and that in its proper colour viz a flesh colour. Hounds tongue hath a forme not much different from its name which will tye the Tongues of Hounds so that they shall not barke at you: if it be laid under the bottomes of ones feet.

Signatures, and therefore a divinely sanctioned usefulness, could be discovered in the lowliest of weeds. The knobbly roots of celandine were signed for piles. The blad-der-shaped seed capsules of shepherd's-purse showed that it was a cure for urinary disorders. Mandrake's append-ages, real or contrived, made it a specific for sterility or failures of passion. Nor was it just the external appear-ances of plants that were dredged for meaning. Their habits were, too. Pellitory-of-the-wall, a widespread weed whose roots can penetrate stonework, would be capable of doing the same with kidney stones and 'the gravel'. The heads of cowslips, which trembled in the wind, were signed for Parkinson's disease, the 'shaking palsy'.

Coles's true weeds, his own plants out of place, were those which had no signature. They seemed at first sight to be on the earth without a purpose. But he warns readers against dismissing them too hastily. Their virtues might simply have not yet been discovered. He goes on to give an explanation for the presence of seemingly useless plants in the world which recalls Virgil's account in *The Georgics* of Jove's creation of weeds. 'They would not be without their use, if they were good for nothing else but to exercise the Industry of Man to weed them out who, had he nothing to struggle with, the fire of his spirit would be half extinguished in the Flesh.'

What is striking, though not of course surprising for the seventeenth century, is the uncompromising human-centredness of the Doctrine of Signatures. Its proponents had no notion that plants might have, so to speak, their own reasons for being the shape and colour they are. The yellow of the dandelion flower could not possibly have anything to do with attracting pollinating insects. It was an indication of the weed's suitability for urinary disor-ders. The hooks on burdock seeds weren't there to help them migrate, but to draw venom from vipers' bites. As for the walnut (a Signaturist's specific for disorders of the head), it was the very archetype of a signed plant. The shape of the shell was designed not so much to perfectly contain the walnut, as to remind us of our brains, and the wisdom God had planted there.

~

The Doctrine of Signatures, extraordinarily, has continued to influence herbal medicine into the modern era. But mainstream plant-based medicine began increasingly to follow the observation-based approach favoured by Turner. The first truly popular English guide to plants and their properties was John Gerard's *Herball, or General Historie of Plants* (1597). Much of it is far from original and based on unacknowledged borrowings from a 1583 text by the Flemish botanist Rembert Dodoens. But Gerard had a passion for plants and a poet's gift for capturing their personalities. He covers some 2,000 varieties of plant in his herbal, and though his accounts of weeds (rarely referred to as such) aren't the earliest in the English language, they are certainly the first to express any kind of aesthetic appreciation. Gerard also had the beginnings of a modern, rational outlook. He despised the more extravagant beliefs of the Signaturists, and was scathing about that touchstone species, the mandrake: 'There hath been many ridiculous tales brought up of this plant, whether of old wives, or some runnagate Surgeons or physicke-mongers I know not . . . but sure some one or more that sought to make themselves famous and skilful above others, were the first brochers of that errour I speake of.' His derision for 'runnagate Surgeons or physicke-mongers' is one of the first broadsides in a war between different views of plant treatments that would transform the practice of medicine in Britain by the end of the seventeenth century. Three professional bodies were at constant loggerheads, and constantly jostling for status: the College of Physicians, an establishment body that tried to regulate medicine across the kingdom; the Society of Apothecaries, the makers and purveyors of medicines; and the Company of Barbers, licensed to perform surgery.

John Gerard was a surgeon himself, though doubtless not a runnagate one. He was apprenticed in 1562, when he was seventeen years old, to Alexander Mason, a London barber-surgeon, and late in his life was elected Master of the Company. But his real passion was gardening, and from 1577 he was in charge of Sir William Cecil's elaborate gardens in the Strand. Gerard's own garden was round the corner in Fetter Lane, and he knew London well. One of the delights of the book is the evocative picture he conjures up of the sixteenth-century capital's plantscape. He describes navelwort growing 'upon Westminster Abbey over the doore that leadeth from Chaucer his tombe to the olde palace'; musk-mallow 'on the left hand of the place of execution called Tyborne [and] among the bushes and hedges as you go from London to a bathing place called Old Foorde'; woody nightshade 'in a ditch side against a garden wall of the right honourable the Earle of Sussex his house in Bermonsey Street'; rue-leaved saxifrage 'upon the bricke wall in Chauncerie Lane'. As the Cambridge Professor of Divinity and botanical historian Charles E. Raven remarked, 'one could forgive many misdeeds in one who can so bedeck the metropolis'.

But Gerard travelled widely round England, and had a wide circle of correspondents. His lavish and precise description of rosebay willowherb, then a rare plant, and of the aerobatic seeds which would make it such a successful weed three centuries later, was based on a plant he obtained from Yorkshire and planted out in his garden: 'The branches come out of the ground in great numbers, growing to a height of six foot, garnished with brave floures of great beauty, consisting of foure leaves a piece, of an orient purple colour, having some threds in the middle of a yellow colour. The cod is long . . . and full of downy matter which flieth away with the winde when the cod is opened.' He'd found a blue pimpernel, similar to those later seen by Edward Salisbury, in 'a chalkie corne field' in Kent, and notes that the flowers of its scarlet cousin were used as a barometer by fieldworkers. They would expect rain the next day if the petals were closed and sun if they were open.

The entries are full of snippets of lore like this. The leggy stems of goosegrass, covered with clinging hooks, were used to strain milk – and also as a remedy for the bites of poisonous spiders. The dried leaves of coltsfoot were smoked like tobacco – 'and mightily prevaileth' against diseases of the lung. The yellow-flowered corn marigold, taken by sufferers from jaundice 'after coming forth of the bath', restored normal skin colour. Gerard's flights of fancy would be exasperating if he weren't such an entertaining writer, and, for all his rants against quacks, he repeatedly endorses cures which sound as if they have emerged from witches' cauldrons. There are sensible recommendations – the tannin-rich leaves of self-heal as a styptic, mint for unsettled stomachs – that are based on long popular experience. But then Gerard's gullibility, or sense of fun, delivers a moment of pure craziness. Writing of cyclamen, he insists that pregnant women should have nothing to do with the plant; they should not even 'stride over the same where it groweth for the naturall attractive vertue therein contained is such, that without controversie they that attempt it . . . shall be delivered before their time'. He had, he assures his readers, fastened a latticework of sticks over the cyclamen in his own garden 'lest any woman should by lamentable experiment finde my words to bee true, by their stepping over the same'.

When Thomas Johnson – Yorkshire gentleman, London apothecary and Royalist soldier – prepared a new edition of *The Herball* in 1633, 'enlarged and amended', he ribbed the author for being 'womanish' in his cyclamen warning, 'led more by vain opinion than by any reason or experience'. He also passed on a morsel of malicious gossip concerning Gerard's surprising record of a wild peony, which he purportedly found on a rabbit warren in Southfleet: 'I have been told that our Author himselfe planted that Peionie there, and afterwards seemed to finde it there by accident: and I do beleeve it was so, because none before or since have ever seen or heard of it growing wild since in any part of this Kingdome.' (A practice which continues in the plant conservation world to this day.)

As slaps on the wrist go this was a gentle one, and Johnson, though more of a pragmatist than Gerard, was far from being a sobersides. One of his responsibilities was conducting apprentices of the Society of Apothecaries on 'herbarising' trips to acquaint them with medicinal plants, and his logbooks of these expeditions are rumbustious reading – as you might expect, given that they recount the adventures of a gang of students let loose on a field trip. The most ambitious outing was to north Kent in 1629. The party of ten set out in two boats from the City of London on 13 July, bound for Gravesend. They were almost immediately hit by a storm and half the party were forced to go ashore at Greenwich. They met up again at Rochester, and spent the night at a hostelry. Drinking seemed to occupy a lot of research time, and it's typical of the mood of the trip that the very first plant in their list of findings was 'A lichen plucked from the wall of the inn'. In the days that followed they meandered through the Kent countryside, towards Chatham and Gillingham, recording an impressive list of plants, in which weeds – just as important in herbal medicine as rarities – had an honourable place. On the first day they noted henbane, hemlock, ragwort, shepherd's-purse, black and woody nightshade, pellitory-of-the-wall (in a cemetery), house-leek, three species of the cornfield weed fluellen, stinking mayweed, shepherd's-needle, yellow-rattle and self-heal. On the Isle of Sheppey they were interrogated by the Mayor of Queenborough – no doubt slightly alarmed by this band of strangers meandering unsteadily about his parish. They satisfied him as to the seriousness of their purpose, and were regaled with Kentish beer. Then they took a barge to the Isle of Grain, and 'walked five or six miles without seeing a single thing that could give us any pleasure'. Or, it seems, a single pub. 'The road ran along the water's edge. In the heat of the day we were tormented like Tantalus with a misery of thirst in the midst of the waters.' What a relief it must have been for the party to find a brewer's dray bound for Rochester. Johnson packed the apprentices on this, waved them off happily 'lolling amongst the barrels' and went on ('lest we be taxed with laziness or carelessness') to discover cannabis and opium poppy at Cliffe.

~

Johnson was created Honorary Doctor of Physic at Oxford in 1643, and one year later, bearing the rank of Lieutenant Colonel in the Royalist forces, was fatally wounded in a Civil War skirmish in Hampshire. In the same year Nicholas Culpeper, the best known if not the most reputable of the seventeenth-century herbalists, was also terribly wounded – fighting for the other side at the battle of Newbury. Their different alliances weren't insignificant. Culpeper was, on the face of it, the kind of quack that Gerard and Johnson railed against. His herbal philosophy was based on abstruse astrological theories and cracker-barrel common sense. But he was a political radical and a populist, and more than any other writer or practitioner in the turbulent years of the English Revolution, attempted to make a system of plant-based medicine available to ordinary people. He never entirely recovered from the chest wound he received while fighting for the Parliamentary Trained Bands, but in the ten years that remained for him wrote one of the most unexpected best-sellers of the century.

*The English Physician* was, so to speak, a tabloid herbal, a cheap, racily written, accessible guide to self-medication with easily available and mostly native plants. Of the some 330 species Culpeper covers, about a third would generally be regarded as weeds today. But *The English Physician* was a weed-guide in a more metaphorical sense, a salute to the idea of commonness, whether it lay in the plants themselves, or the manner of their use. Its vernacular style buzzes like a swarm of wasps, but is oddly beguiling. By turns it is obsessive, charming, impenetrable, comforting, satirical, combative and monomanic. The 'furious biting Herb' crowfoot (probably creeping buttercup), the author snorts, is 'very common everywhere, unless you run your head into a Hedge'. A thoroughgoing rational will find the book's excursions into superstition and cosmological fantasy preposterous. Even Culpeper's contemporaries thought it was the work of a self-seeking idiot. But his fancifulness and occasional quackish ravings are at least partially balanced by the motivation behind the book. This is a *people's* herbal, addressed to the suffering poor themselves. It contains no Latin, no dangerously toxic plants, and few species that can't be found in an ordinary garden or country lane. It was, one biographer has gingerly suggested, the first step in the long journey towards the National Health Service.

Culpeper was born in Ockley, Sussex in 1616, just a fortnight after his father had died, possibly of typhoid. He was brought up in the house of his grandfather, the Revd William Attersoll, rector of the village of Isfield, forty miles away. As the only child in the household, he would have had more than his share of the company of women, and their medicinal and culinary arts. He also had, on his doorstep, the great green waste of Ashdown Forest, the setting, three centuries later, of Winnie the Pooh's Hundred Acre Wood, and he doubtless frolicked there like A. A. Milne's characters. His grandfather sent him to Cambridge in 1632, when he was sixteen years old. He was intended to study theology, but whatever ambitions he or his family may have had were abruptly terminated a couple of years later. Nicholas had fallen in love with a Sussex girl who was far above him in social status. It looked like a hopeless liaison, and with an impetuousness he was to show throughout his life, he persuaded her to elope with him. They planned to marry in a chapel in Lewes. Nicholas set out from the rectory in Isfield by coach, his *femme fatale* on foot from the Big House nearby. Close to their rendezvous Nicholas received the stunning news that his bride-to-be had been struck dead by lightning as she crossed the South Downs.

The catastrophe had a profound impact on Culpeper's outlook and future life. He was already beginning to fall under the spell of astrology and religious radicalism, and this bolt from the blue must have seemed both incomprehensibly unfair and a kind of punishment for his rash scheme. Thereafter he had the look of a man with a mission, to discover ways of healing which challenged the arbitrariness of disease and natural disaster. Within days of his fiancée's death he'd abandoned Cambridge and set off for London. He became an apprentice at Simon White's apothecary's shop at Temple Bar, and amongst the legacy of herbal lore he began to absorb, he would have learned about 'Houseleek, or Sengreen', which can 'preserve what it grows upon from fire and lightning'. As a result of this apprenticeship, he found himself elected into the Society of Apothecaries, and set up – as astrologer and herbalist – at his own shop in Spitalfields, east of the City.

Revolutionary London at this moment was a ferment of extreme and eccentric cults and ideologies – all making the most of this brief window of civil chaos: religious

sects, moon worshippers, political subversives, underground publishers, and outright wizards, like Arthur Dee,
son of Elizabeth I's infamous magus, Dr John Dee. Every
kind of would-be healer flourished, given additional fertile ground by a series of raging epidemics and failed harvests. There was a feeling of apocalypse in the air.

The turmoil was reflected in the more orthodox medical world. William Harvey had turned accepted thinking
about the human body upside down by revealing the cir-
culation of the blood and the true function of the heart.
Harvey was the Royal Physician and the commanding figure in the College of Physicians, the establishment body
that was locked in a simmering demarcation dispute with
the Apothecaries and the Company of Barbers. The College was attempting to restrict the activities of apothecaries to the blending and dispensing of medication, and to
forbid them from recommending treatments or selling
'curious' remedies. In April 1618 it had obtained a royal
proclamation to order all apothecaries to buy and adhere
to the newly published *Pharmacopoeia Londinensis*, which
gave an exhaustive list of the remedies they were permitted to produce, and the ways this should be done, down
to the last drachm and distilled drop.

The *Pharmacopoeia* seemed on the surface an enlight-
ened attempt at regulation, a way of keeping the worst
excesses of unlicensed apothecaries in check. But in reality it was just an extension of the College's project to con-
solidate its overarching authority. Anyone who believes
that Culpeper was a quack by the standards of his time as
well as our own should consider the *Pharmacopoeia*'s list of
approved substances. Ranging alphabetically from ants to
wolf's intestines, it takes in horn of unicorn, human tears,

pounded swallows, moths, vulture fat, bozoar stones from the intestines of Persian wild goats as well as toxic exotic herbs and some plants that were entirely imaginary.

Culpeper launched himself into this fractious world like a trader in an East End market. He gave consultations from his shop and soapboxed his products and beliefs on the street. Garden historian Eleanour Sinclair Rohde has conjured up his charisma: 'It is impossible to read any part of [his] absurd book without a vision arising of the old rogue [young, actually], standing at the street corner and not only collecting but holding an interested crowd of the common folk by the sort of arguments which they not only understand but appreciate.' He lambasted Classical writers, trumpeted his own superior knowledge, even denounced other varieties of astrologer. Those who cast horoscopes over the contents of their patients' chamberpots he dubbed 'piss prophets'. Most of the rest – the 'hororary' astrologers, who prescribed remedies according to the moment the patient reported the complaint – he declared 'as full of nonsense and contradiction as an Egg is full of meat'. His own astrological technique was called 'decumbiture', and required the drawing up of a chart for the exact moment the disease manifested itself or the injury occurred. This was considered the moment when planetary and cosmological forces most powerfully converged in the patient's body. Then it was a matter of choosing as a remedy a herb which lay under the same astrological conjunctions – a runic process which Culpeper never explains. But plotting the charts did mean taking a thorough history of the illness, which places him, in modern parlance, as a holistic practitioner, concerned about the patient as well as the disease.

Whatever its bizarre ideological roots, Culpeper's approach went down famously in Spitalfields and its environs. He became popular with local people for his cheap fees and no-holds-barred irreverence. An eye-witness sketch, published a century and a half later in the *Gentleman's Magazine* of May 1797, talks of his long brown hair and 'spare lean body' and reveals that he was 'a good orator though very conceited and full of jest', none of which can have done his image any harm.

He was beginning to make an impression in the wider world, too. He lectured to the newly formed Society of Astrologers. He published a stream of pamphlets and books, on the significance of a solar eclipse occurring close to the beheading of King Charles, on midwifery, on domestic medicine. But – a paradoxical Roundhead rake to the core – he ran into trouble as well. Just after his marriage in 1640, he fought a duel. The reason isn't known, but Culpeper had to pay for his opponent's treatment, and then hide away in France for three months. In 1642, with England on the brink of Civil War, he was convicted of witchcraft. A woman called Sarah Lynge had been treated by him, and promptly started to 'waste away'. She accused him of Devil's dealing, and Culpeper was arrested and tried. He was gaoled in 1643, probably in Newgate. Shortly after his release he joined the great throng of the Trained Bands as they marched out of London to relieve their comrades besieged at Gloucester (ironically by a remote Royalist relative of Nicholas, Sir John Culpeper). On the way they encountered the Earl of Essex's troops at Newbury, and a horrendous three-hour battle ensued, in which Nicholas Culpeper was struck in the chest by shrapnel.

He never entirely recovered from his wound, and was to die just ten years later, at the age of thirty-seven. But the trauma seemed to have a steeling effect on his purpose, just as had that earlier *coup* on the Downs. Sometime around 1648 he accepted a commission from the radical publisher Peter Cole to prepare an English translation of the latest edition of the *Pharmacopoeia*. It seemed, to the medical establishment, an act that was close to treason, or maybe blasphemy. The secrets that the College had clasped so close to its chest were to be broadcast about like weed seeds, the privilege of healing made accessible to Everyman. Culpeper's *A Physical Directory* was published in August 1649. The outrage it caused to the College of Physicians was compounded by the fact that Culpeper had not merely made a translation, but added whole new sections of his own, including the uses of a hundred plants not in the original edition. Culpeper made it clear that his courageous, possibly foolhardy interpretation was a political as much as a medical statement. The very first page is a signed letter from 'The Translator to the Reader', an extraordinary declamation which links the bodies politic and human:

> The Prize which We now, and They within a few years shall play for, is, THE LIBERTY OF THE SUBJECT . . . So far as I can see by the help of my Optic Nerves (whether it be the *Intromittendo Species* or *Extramittendo Radios*, it matters not much) the Liberty of our *Common-Wealth* (if I may call it so without a Solecism) is most infringed by three sorts of men, *Priests, Physicians, Lawyers* . . . The one deceives men in matters belonging to their Souls, the other in matters belonging to their Bodies, the third in Matters belonging to their Estates.

The book was an instant success, and within weeks pirated editions began to appear. Culpeper, rising in confidence, decided to clasp the moment, and in the 1651 edition of his *Directory* advertised that he would publish a popular guide 'in the knowledge of Herbs before I am half a year older'. His herbal duly appeared that autumn, under the title *The English Physitian*, and the subtitle: 'A Compleat Method of Physick, whereby a man preserve his Body in Health, or cure himself, being sick, for three pence charge, with such things as grow only in England'.

Most previous herbals had been lavish productions for the intelligentsia. Even Gerard's sympathetic and readable volume was in reality an expensive parlour-table production intended for the gardening middle class, not the typhoid-infested poor. Culpeper's innovative book broke with convention. It was priced, as he promised, at threepence. The plants are arranged not according to some arcane taxonomy, but in rough alphabetical order, from adder's-tongue to yarrow. Almost all the species included are English natives, save for a few, such as artichoke and walnut, which were frequent enough in gardens. (This was a pragmatic decision, to ensure that they were easily and cheaply available, not an expression of the more mystical belief of later herbalists that English plants growing in English soil were more suited to English bodies.) Each plant is described, as is its habitat, its time of flowering and its virtues and uses. Each is given a 'governing' planet: ground-ivy is under Venus, chickweed under the Moon, nettles under Mars. The rationale behind these associations is never explained, though there are occasional hints. Under 'Arssmart' (water-pepper) he suggests, gnomically, 'that which is hot and biting, is under the dominion of

Mars, but Saturn challengeth the other, as appears by that leaden coloured spot he hath placed upon the leaf'.

Yet despite his passionate defence of the importance of astrological influences, they don't make much of an appearance beyond these initial labellings. I suspect Culpeper hoped they would act as a kind of *mana*, the seventeenth-century equivalent of the modern advertiser's 'Ingredient X': buzz-words designed to inspire respect and recall. For the most part the entries on 'virtues and uses' are an eclectic brew of Signaturist dogma, unacknowledged references to Classical prescriptions, and a good measure of old-fashioned country lore.

Dandelion is beautifully described, its florets 'nicked in with deep spots of yellow in the middle', and is said to be 'under the dominion of Jupiter'. But the account of its action in opening 'the passages of the urine both in young and old', and the mention of its country name of 'Piss-a-Bed' owe nothing to astrology. Nor did his joshing, street-trader's recommendation of lesser celandine for piles, a Signaturist's nostrum justified by the similarity between the plants knobbly roots and swollen veins: 'Here's another secret for my countrymen and women . . . Pile-wort made into an oil ointment or plaster, readily cures the piles or haemorrhoids, and the king's evil [scrofula] if I may lawfully call it the king's evil now there is no king . . .' Docks are recommended as good for the kitchen as well as the surgery, 'as wholesome a pot-herb as any grows in a garden, yet such is the nicety of our times forsooth, that women will not put it in the pot because it makes the pottage black: pride and ignorance (a couple of monsters in the creation) preferring nicety before health'.

The key to understanding the book, by Culpeper's own

insistence, is wormwood, a pungent silver-leaved perennial weed later to become notorious as the raw essence of absinthe. But the entry on this species is unlike anything else in the herbal. It reads like the ramblings of a drunk, an incomprehensible prattle that takes in a three-part conversation between Mars, Venus and himself, and a story about wormwood deterring moths because they are both governed by Mars. Perhaps Culpeper was on the wormwood himself when he wrote it, under the influence not of Mars but of the hallucinogenic chemicals that give absinthe its kick. 'Melancholy men', he writes immediately afterwards, 'cannot endure to be wronged in point of good fame.' Perhaps, suggests Benjamin Woolley in his biography of Culpeper, this stream-of-consciousness entry on one of the bitterest of herbs is an allegory about bitterness itself, the deranged reflections of a man whose life had been a constant battle with the Establishment, and who was close to dying because of a sacrifice made while defending his alternative view of the proper order of things.

Culpeper's influence lives on. There is a chain of Culpeper shops, selling herbs – though a considerably smaller and blander range than described by the eponymous author, and stronger on herb-garden plants than wayside weeds. His book is still in demand, but read, I suspect, as a quaint guide to an Olde Worlde, and not as one of the stranger extrusions of an urban England ripped apart by revolution. It is certainly no longer widely used as a self-treatment guide. Yet the plants it recommends are unlikely to have done anyone any harm. Several may well have helped with minor ailments, though perhaps not for the reasons Culpeper suggests.

His most important legacy may be more subtle and indirect. In his insistence on simple, non-toxic medicaments, and on the public's right of access to pharmaceutical information, he added his mite to the democratisation of medicine, and helped guarantee that outlandish and objectionable substances would soon disappear from the pharmacopoeias. Even a nettle poultice (to refresh 'our wearied members') seems preferable to pounded swallows.

# 6

## *Love-in-idleness*

### Three writer's weeds

WILD PANSIES ARE WIDESPREAD cornfield weeds and are separated botanically into two main species. Wild pansy, *Viola tricolor*, or heartsease, with flowers in combinations of violet and yellow, is the fussier, occurring thinly on sandy and acidic soils throughout Britain. The smaller-flowered field pansy, *Viola arvensis*, is common in cultivated ground everywhere. They are both highly variable in size and colouring, and hybridise freely where they grow together.

Despite their commonness and intriguing looks, pansies weren't much used in herbal medicine. Gerard thought they were useful for treating convulsions in children, itching and venereal disease. Culpeper agreed, adding, with a characteristic leer: 'The herb is really Saturnine; something cold, viscous and slimy. A strong decoction of the herbs and flowers . . . is an excellent cure for the French pox, the herb being a gallant antivenereal.' This suggestion rather goes against the reputation heartsease had outside medicine – or perhaps was an instance of magic's frequent and often homeopathic balancing acts: the weed that provoked a disease being the best one to cure it – because in the commonplace world pansies were love tokens and fancies. From at least the Middle Ages, they enchanted people, stirred up romantic imaginings. In this they are one more piece of evidence against the conventional wisdom that country people were too busy or too stupid to have anything other than a doggedly practical interest in wild plants.

It's not hard to see why. The pansy's flowers are composed like a face. They have two high brows, two cheeks and a chin, and streaks that suggest eyes, or laughter lines. They are normally a pale cream with a few purple striations, but individual flowers are marvellously unpredictable, as if they've been randomly daubed with a watercolour brush. There can be dark eye patches, and purple beauty spots on the brow or chin. I've found flowers which are striped or speckled with blues and violets, and a few that are wholly purple.

In France the meditative 'face' suggested a thinker, and in the Middle Ages the flowers were known as *pensées* (thoughts), later Anglicised to 'pansy'. But in English parishes people saw two faces, up to much less intellectual business. They were kissing, the side petals lip to lip within a hood formed by the upper petals. Kiss-and-look-up was the nickname in Somerset, and elsewhere, Kiss-behind-the-garden-gate, Kiss-me-over-the-garden-gate, Kiss-me-quick, Leap-up-and-kiss-me, climaxing in Lincolnshire's upstairs-downstairs version, Meet-her-in-the-entry-kiss-her-in-the-buttery. But the wild pansy was more widely known as heartsease, and perhaps it was used for just that purpose, picked as a posy to claim the kiss it pictured.

There was a more melancholy name in Warwickshire and the west Midlands. Love-in-idleness emerged perhaps because the pansy's three lower petals could be seen as a woman flanked by two lovers; a flower therefore representing frustrated, fruitless, 'idle' love. And it was this sense that Warwickshire's brightest son latched on to in the late sixteenth century, and spun into an extravagant poetic fantasy about the plant.

Shakespeare's *A Midsummer Night's Dream* is probably the only play in the English language whose plot hinges on the potency of a weed. The mayhem and mistaken identities in the forest come about because Puck, the King of the Fairies' odd-job man, squeezes the juice of love-in-idleness onto the characters' eyes while they are asleep. They will fall in love with the first creature they set eyes on when they wake.

Shakespeare was born and bred in Stratford-upon-Avon, and knew the wild flowers and folklore of Warwickshire intimately. And he takes it for granted that his audience will also be familiar with them, and with all their vernacular names and vulgar associations. More than a hundred species of wild plant are mentioned in his works, and it's no surprise that a good number of them are those commonplace plants, so redolent with meaning, the weeds.

The common daisy, the 'pied' daisy of *Love's Labour's Lost*, crops up in no fewer than four of the plays, and in *The Rape of Lucrece* is a metaphor not just for virginal whiteness but the arrival of spring:

> Without the bed her other fair hand was
> On the green coverlet, whose perfect white
> Showed like an April daisy on the grass

The daisy is also part of the drowning Ophelia's 'fantastic garland': 'crow flowers, nettles, daisies and long purples', the identities of which are still argued over by botanists and critics. Shakespeare's own audience would have known very well what they were, and what they symbolised. Natural metaphors were common currency in the sixteenth century, and Shakespeare deploys them in a constant stream of puns, allusions, winks and nods, many of which are so topical or local that they've lost their resonance. The elegiac lines from *Cymbeline* – 'Golden lads and girls all must / As chimney sweepers, come to dust' – seem an odd metaphor unless you're aware that 'chimney sweepers' was Warwickshire patois for the wind-scattered, time-telling 'clocks' that follow dandelion's golden flowers.

*A Midsummer Night's Dream* crackles with plant imagery. Most of the play is set in a forest, notionally near Athens, though clearly an English landscape growing English flowers. But the location isn't literal. The botanical *dramatis personae* are from different seasons and different habitats. Even in Warwickshire's Forest of Arden you couldn't at one moment of the year assemble a bouquet of pansies together with the luscious but bewildering floral ingredients of Titania's bank, 'whereon the wild thyme grows'.

The *Dream*'s plot is deceptively simple. A grand wedding is planned by Egeus, an Athenian nobleman, to join his daughter Hermia with Demetrius. But she refuses to go through with the marriage because she loves another man, Lysander. She flees to the forest, pursued by her intended and her best friend Helena, who secretly adores Demetrius. But there is trouble in the woods already. Oberon, King of the Fairies, has quarrelled with his

queen, Titania, because she refuses to give him a change-
ling boy (stolen by her fairies) as a page. Then comes the
weed spell, a bit of botanical comic business that turns the
action into pure farce.

It is part of Shakespeare's genius that he is able to transform his knowledge – of plant lore, for instance – into fantastical dramatic devices. It's a skill he would have begun
to learn at school, the art of what the Elizabethans called
'lively turning'. You took a superstition, a scandal, a myth,
a real historical incident, and with a deft narrative twist,
gave it new life. Puck, Oberon's familiar, is also a lively
turner. He's a version of Robin Goodfellow, mischievous
and hedge-wise. When Oberon, smarting from Titania's
obstinacy, orders him to fetch the juice of a special flower
and squeeze it on her eyes while she is asleep, so that she
will 'madly dote' on the first creature she sees, Puck gets
carried away by his own impish enthusiasm, and drizzles
the magic juice onto most of the frustrated lovers roaming
the wood.

Shakespeare here is mixing Classical myth, Midlands
vernacular and sheer comic invention. Oberon describes
the pansy as 'a little western flower', bringing it back from
the remoteness of Athens to his audience's home fields.
But it has been enchanted by one of Cupid's arrows, which
changed its colour from milk white to 'purple with love's
wounds' – a description which faithfully reflects the colour
variations of heartsease as well as echoing a story in Ovid
about the mulberry changing from white to the colour of
a bloodstained veil. He calls the flower by its most redolent local name – love-in-idleness – a plant tailor-made for
the young Athenians' amorous agonies. But the device of
Puck squeezing its juice onto the hapless characters has no

roots whatsoever in folklore, and I imagine Shakespeare simply invented it, as a useful comic device.

~

That's probably as far as I would have got into Shake-speare's plant symbolism by myself. But I was lucky enough to be able to experience a professional's take on the subject. In 2005 Greg Doran, Director of the Royal Shakespeare Company at Stratford, was mounting a new production of the *Dream*, and invited me to join him in an exploration of the play's use of natural symbols for an accompanying television documentary. He was espe-cially intrigued by the description of Titania's bank, and what gave this botanical incantation its extraordinary charge:

> I know a bank whereon the wild thyme blows,
> Where oxlips and the nodding violet grows,
> Quite o'er-canopied with luscious woodbine,
> With sweet musk-roses, and with eglantine.
> There sleeps Titania sometime of the night,
> Lull'd in these flowers with dances and delights.

It is certainly a very odd list. The plants are not exactly weeds, though they are all wild (except for one, the musk-rose). But their superficial differences overshadow their similarities. They are bushes and climbers and tiny tufting perennials. They grow in different habitats, and flower at different times of the year.

This proved a challenge not just to disentangling the meaning of the lines, but to the logistical planning of the programme, as Greg wanted to film the discussion on location. We mulled over different sites, weighed travel time against scenic bonuses, peered at long-range weather forecasts, and eventually settled for a spectacular chalk hill in the Chilterns which I knew moderately well, and where I reckoned we could score four of Titania's six species. We climbed towards Turville's windmill just a couple of days before Midsummer. The oxlips (false) and violets (sweet) were long blown, but we did find 'luscious eglantine' (sweet-briar) and a bona fide bank of wild thyme.

We sat on the bank and gazed down at the village in the valley, and pondered Titania's evocative litany. Red kites and buzzards – newly returned to these hills – wheeled in the thermals, just as they would have done in Shakespeare's day. Below us the wheatfields, rimed with white chalk, seemed to be smouldering with the glow from immense patches of vermilion fumitory. The weed is named from its wispy, grey-green leaves, fancifully thought to resemble mist: *fumus terrae*, smoke of the earth. But here, in full flower, it looked more like embers of the earth. Greg reminded me of Shakespeare's use of its vernacular name 'fumiter' in describing mad King Lear's wreath: 'singing aloud / Crowned with rank fumiter and furrow-weeds, / With hardocks [probably burdock], hemlock, nettles, cuckoo-flowers, / Darnel, and all the idle weeds that grow/ In our sustaining corn'. It was Lear's ultimate sign of disorientation, to elevate weeds into a crown. Listening to Greg intone the lines I could register the power of the plants' names, rattling like insults. He told me about the *Dream*'s origins, how it was written in honour of the wedding of one of the playwright's patrons and is full of private and topical jokes. One of Puck's fairy friends sings about cowslips: 'In their gold coats spots you see. / Those be rubies, fairy favours / In their freckles live their savours'. She calls them 'pensioners' after the inner circle of Queen Elizabeth's courtiers who flounced about in extravagant costumes covered with gold embroidery.

We go through the list of Titania's flowers, and it seems to me that the only link is a high perfume. Wild thyme is spicy, and Gerard's *Herball*, published in 1597, the year after the *Dream*'s first performance, describes it as 'aromaticall'. Violets are amongst the sweetest smelling of all wild flowers, and Shakespeare refers to them often. In *The Winter's Tale* they are 'sweeter than the lids of Juno's eyes / Or Cytherea's breath'. Woodbine is honeysuckle, whose flowers are most powerfully scented at night. Eglantine's (sweet briar) leaves smell enchantingly of apple, especially after rain. Musk-rose speaks for itself. None of these plants was literally aphrodisiac – what Elizabethan herbalists called 'venereous' – but their seductive scents were more likely to stir Titania up than 'lull' her. She was chasing 'dances and delights' more than a good night's sleep.

Only the oxlip seems out of place, being neither fragrant nor symbolic. This isn't the botanist's oxlip (*Primula elatior*, confined to East Anglia and unrecognised in Shakespeare's day) but the widespread hybrid between the primrose and the cowslip. Greg suspects that the reference might be a coded joke, perhaps a nickname for the poet's patron, or a much cruder reference to his betrothed. I wonder if 'oxlip' is there simply to add to the wonderful belling of l's in the list: wild, violet, luscious, eglantine, lull'd; the mellifluous l's of love and lust, and, at the end of the stanza, the erotic image of a shed snakeskin, lying 'enamell'd' on the bank. In a similar but, as it were, converse way, King Lear's list of troubling weeds – 'hardocks, hemlock, nettles, cuckoo-flowers' – crackles with cussed k's. His wreath would sound uncomfortable and cranky, and Titania's bed seductive, even if we knew nothing whatsoever about the plants. Both lists are theatrical *spells*, designed to beguile or disgust the listener by their sound as much as their literal content.

In all his writings, of course, Shakespeare's language is multi-layered: descriptive, allusive, sonorous all at once. His confident use of weeds as symbols suggests that their popular meanings aren't (or at least weren't) superficial, concerned purely with agricultural nuisance, but have cultural and ecological undertones that are built into the genetic structure of their names.

～

Two centuries later the poet John Clare took an alternative stand on the naming of the wild pansy. In 1820 he had published his first collection of verse, *Poems Descriptive of Rural Life and Scenery*, and caused something of a stir, chiefly because he was a young Northamptonshire rural labourer who wasn't afraid to use local patois. The poems contain some startlingly vivid and intimate writing about wild flowers and weeds, in which Clare – in an unprecedented shift of perspective – salutes them as fellow citizens. 'Welcome, old matey!' begins the first stanza of 'To an April Daisy', 'Hail, beauty's gem! Disdaining time nor place / Carelessly creeping on the dunghill's side'.

The *Poems* took the fancy of Elizabeth Kent, a gardening writer and sister-in-law to the essayist Leigh Hunt, and she mentioned them in her *Flora Domestica, or the Portable Flower-garden*, published by Taylor and Hessey in 1823.

They were Clare's publishers too, and they sent the poet a complimentary copy. Clare enjoyed its discursive mix of down-to-earth gardening lore, lyrical plant description and poetical allusions, and doubtless (despite her patronising tone) Kent's tribute to his own flower poetry: 'None have better understood the language of flowers than the simple-minded peasant-poet, Clare, whose volumes are like a beautiful country, diversified with woods, meadows, heaths and flower-gardens.' Shortly afterwards he wrote back to Hessey with some notes on his local flora that he thought might eventually be worked up into a book, and which reflect an altogether more direct view of the relationship between wild flowers and humans than Shakespeare. '"*Hearts-ease*"', he writes in the letter (between '*Guelder Rose*' and '*Heath*'),

> they are calld with us 'panseys' & 'pinking-Johns' but for what reasons I cannot tell there is a wild sort in our fields with a very small yellow flower & leaves exactly like the garden one I have tryd to cultivate it to see what it woud change to but it lovd its wild state so well that it was too stubborn for me & I gave it up to its fields agen – how much I love some of the names that this author [Kent] has picked up what a characteristic name for the hearts ease is the 'butterfly violet' & the 'wingd violet' the first is the best – I don't like L. Hunts 'Sparkler' it's a consciet rather fitting a name for liquors th[a]n flowers.

Did Clare also consider 'love-in-idleness' a 'consciet', see it as a little too mannered and urbane? It was a name current in his native Northamptonshire but he never

mentions it. From the outset his passion for plants is
focused on their vitality and independence. He writes
of them as fellow creatures, with their own life-plans and
dwelling places. His poems are full of subtle allusions to
their vernacular uses and literary associations, but first
and foremost he writes of them as things-in-themselves,
not as a colourful palette of symbols and metaphors. In
this he differs fundamentally from Shakespeare, and he is
not afraid to be explicit about this difference, even though
he knew and admired the playwright's work. Early on in
his 1824 flower letter, he argues about the identity of the
plant which has the vernacular name 'cuckoo':

> '*Cardamine*' . . . this is called lilac with us as well
> as 'ladysmock' but I never heard it called cuckoo
> in my life otherwise th[a]n by books – the wood
> anemone is also called 'lady smock' by children
> what the common people call 'cuckoo' with us is
> one which is a species of the 'Orchis' . . . these
> is my cuckoos & the one that is found in Spring
> with the blue bells is the 'pouch lipd cuckoo bud'.
> I have so often mentioned its flowers are purple
> & freckld with paler spots inside & its leaves are
> spotted with jet like the arum they come & go with
> the cuckoo & in my opinion are the only cuckoo
> flowers of England let the commentators of Shak-
> spear say what they will nay shakspear himself has
> no authority for me in this particular the vulgar
> wereever I have been know them by this name
> only & the vulgar are always the best glossary to
> such things.

For Clare 'vulgarity' was a touchstone of worth and

authenticity. It was a concept that embraced commonness, lowliness and a lack of pretension, qualities he admired in both humans and nature. Later in her tribute to him, Elizabeth Kent wrote (more respectfully) that 'This poet is truly a lover of Nature: in her humblest attire she is still pleasing to him, and the sight of a simple weed seems to him a source of delight.' Clare is rarely openly anthropomorphic, but weeds for him are the poor peasants of the plant world – ubiquitous, modestly handsome, ignored; useful, yet constantly abused; troublemakers, yes, but really just attempting to live their lives as best they can. In what is probably the most extended passage on weeds in English poetry (in the title poem of *The Shepherd's Calendar*, 1827), he describes the progress of the weeding gangs (still using the same tools as described by Tusser 300 years before) and how they engaged with the plants they were demolishing:

> Each morning now the weeders meet
> To cut the thistle from the wheat
> And ruin in the sunny hours
> Full many wild weeds of their flowers:
> Corn popies that in crimson dwell
> Called 'headaches' from their sickly smell;
> And carlock yellow as the sun
> That o'er the May fields thickly run;
> And 'iron weed' content to share
> The meanest spot that souring can spare –
> E'en roads where danger hourly comes –
> Is not wi'out its purple blooms
> And leaves wi' pricks like thistles round
> Thick set, that have no strength to wound,
> That shrink to childhood's eager hold

Like hair; and with its eye of gold
And scarlet starry points of flowers
Pimpernel, dreading night and showers,
Oft called 'the shepherd's weather glass',
That sleeps till suns have dried the grass
Then wakes and spreads its creeping bloom
Then close it shuts to sleep again.
Which weeders see and talk of rain
And boys that mark them shut so soon
Will call them 'John go to bed at noon';
And fumitory too, a name
That superstition holds to fame
Whose red and purple mottled flowers
Are cropped by maids in weeding hours
To boil in water milk and whey
For washes on an holiday
To make their beauty fair and sleek
And scour the tan from summer's cheek;
And simple small forget-me-not
Eyed wi a pin's-head yellow spot
I' the middle of its tender blue
That gains from poets notice due.
These flowers their toil by crowds destroys
And robs them of their lonely joys,
That met the May wi' hopes as sweet
As those her suns in gardens meet;
And oft the dame will feel inclined
As childhood memory comes to mind
To turn her hook away and spare
The blooms it loved to gather there.

Molly Mahood, in *The Poet as Botanist*, has pointed out
how *garrulous* this scene sounds. Clare may have been the

only person present with a conscious fascination with language, but the whole weeding gang are gossiping: the boys, the maids, even the 'dame' (weeding knew no barriers of sex and age), pointing, chatting, maybe slipping the odd magical herb such as fumitory into their apron pockets. For Clare the weeds too are partners in the conversation. He feels a kind of solidarity with them, as fellow members of the great commonwealth of the fields. Coming across shepherd's-purse in an unfamiliar place, he calls it 'an ancient neighbour . . . Its every trifle makes it dear'. Of the common daisy he says, 'the little daisey wears the self same golden eye and silver rim with its delicate blushing stains underneath in our fenny flats as it does on the mountains of Switzerland if it grows there'.

We get no sense that farm labour is being glamorised by Clare's colourful description because we know this is first-hand experience, not sentimental observation from the field gate. Clare was one of the weeders himself, a participant in what is graphically known as 'stoop labour' in America. He knew weeding was necessary to control the 'destroying beauty', and would likely have agreed with the poet and critic Geoffrey Grigson (an early champion of Clare) when he said, 'When I see men, and women, bent over the crops, I realise it isn't so agreeable for them.' But for Clare the stoop was also an habitual, voluntary reflex, a way of getting closer to the earth. He talks often of 'dropping down' to peer more closely at a plant or insect – or to scribble the first drafts of his poems on the back of old seed packets and sugar bags. It was a bird-like move, inquisitive and impulsive, but also deeply sympathetic. Clare wants to be part of the community of the soil and to look at the world from, so to speak, its own point of view. As in Dürer's *Large Piece of Turf*, this transforms the significance – and scale – of the modest organisms in his rhapsodic gaze. 'On Sundays', he writes in a journal note, 'I used to feel a pleasure to hide in the woods instead of going to church to nestle among the leaves and lye upon a mossy bank w[h]ere the fir like fern its under forest keeps "In a strange stillness" watching for hours the little insects climb up and down the tall stems of the woodgrass and broad leaves.' A patch of weeds becomes a forest. A 'huge keck' (a hogweed, probably) has the majesty of 'a timber tree'. A spread of furze in flower beneath a rainbow is transformed into 'a golden ocean'. In a few of his verses addressed to weeds the title itself is stretched as if to compensate for the modesty of the plant: 'To an Insignificant Flower Obscurely Blooming in a Lonely Wild'; 'An Anniversary, To a Flower of the Desert'.

What happens from this intense and wilfully close-focused gaze is that, in art historian Elizabeth Helsinger's words, 'seeing becomes *heeding*'. Anything that seems significant for the plant – the protective curl of a stem, the speckling on a petal – is significant for Clare, too. His descriptions are minutely exact – '[c]rimp frilled daisy, bright bronze buttercup' – but also convey a real ecological sense that even the most obscure 'weedling wild' has biological connections with its habitat and with all the other creatures that live there. (The 'weedling wild' is the subject of a ballad about dignity – Clare's and the weed's – and ends with Clare deciding not to 'crop its tender flower' but to take the whole plant, root, soil and all, back to his garden to grow it on in safety.) In 'The Lament of Swordy Well' he takes dropping-down to its logical – and ecological – conclusion, and writes as if he were 'a piece of land'. Swordy Well was a patch of grassy common on the southern edge of Helpston that had been repeatedly abused, grubbed up for wheat, quarried for sand, stripped for turves. Clare gives it a voice, to lament its fate and describe the complex network of life that once depended on it. Weeds were part of the system, helping to hold the soil in place and providing food for insects: 'The butterfly's may wir and come / I cannot keep 'em now'.

In a sonnet to 'The Ragwort' he places the weed exactly in its habitat and season:

> Decking rude spots with beautys manifold
> That without thee were dreary to behold
> Sunburnt & bare – the meadow bank the baulk
> That leads a waggonway through the mellow
>     field

Clare's quiet praise of ragwort perhaps shows most clearly how far we have moved from an ecological understanding of weeds. Ragwort is currently regarded as one of the most noxious of all native plants. It contains alkaloids which, if ingested in large quantities by grazing animals, cause irreversible liver damage. They die with distressing symptoms, including the chaotic muscle twitching known as 'the staggers'. Ragwort currently accounts for up to half of all poisoning cases amongst farm animals. It is included in the Weeds Act of 1959, and the targeted Ragwort Control Act of 2003, which require landowners to take action to prevent its spread. The owners of horses (frequent victims) regard it as an 'epidemic' that needs to be expunged from the countryside by any means, including blanket spraying.

Yet on closer inspection the situation is more complex. Neither wild nor domestic animals will usually eat growing ragwort if other forage is available. The vast majority of poisoning cases are from dried plants which have been cut with hay, and, ironically, from wilted and shrunk specimens which have been sprayed with herbicide (the plant is just as toxic when dead, and less easily recognised by animals). Nor does ragwort seem to have been a particular problem in Clare's day or at any previous time. Its effects were known, but I haven't been able to find a mention of ragwort in any early farming manual. Vernacular names are usually a reliable record as to how plants were viewed, and ragwort's refer either to its appearance (Yallers, Yellow weed), its rank smell (Stinking Billy, Mare fart), or its time of flowering (Summer farewell, James's weed). There is just one rarely used vernacular name, Staggerwort, which refers to its effect on cattle.

Clare's feelings about flowers were exceptional, but he was a fieldworker himself, and would not have written so lyrically of the plant if it disabled animals as it does today. Was ragwort less common then (unlikely), or more sensibly managed, or simply kept at a distance, and shown a sensible respect? Were farm animals less pampered and wiser in their grazing choices? Whatever the reasons, Clare accepts ragwort as one of the adornments of the summer landscape, even by the side of the 'waggonways' used by horses. The absence from the poem of any reference to local hostility (often mentioned in connection with other species) suggests there was some kind of rapprochement with the plant. It was a weed to be respected, not demonised.

> . . . & everywhere I walk
> Thy waste of shining blossoms richly shields
> The sun tanned sward in splendid hues that burn
> So bright & glaring that the very light
> Of the rich sunshine doth to paleness turn

There is a long road between that sentiment and the 2003 Ragwort Control Act. Times change, and what gives many of Clare's weed poems their intense charge is that they are elegies, memorials to a floral landscape that had been violated and from which humans were to become increasingly alienated. In 1809, when he was sixteen years old, an Act of Parliament was passed for the enclosure of Helpston and its four surrounding parishes – Clare's 'whole world'. Over the next eleven years the matrix of habitats that he knew so intimately was transformed. The big open fields were broken up and fenced and the various fragments distributed amongst the private landowners. Streams were stopped-off so that the new drainage ditches could be straightened. Roads too were straightened or blocked, old trees destroyed, and the first 'No Trespassing' signs appeared. Most painful for Clare (who had, ironically, helped with the fencing of the new fields) was the ploughing-up of the commons and heaths where he'd roamed since he was a boy. In 1821, the year after the enclosure was completed, he published his second collection of poems, *The Village Minstrel*, and in the title poem vents his rage against the dying of the weeds:

> There once were springs, where daisies' silver studs
> Like sheets of snow on every pasture spread;
> There once were summers, when the crow-flower
>     buds

Like golden sunbeams brightest lustre shed;
And trees grew once that sheltered Lubin's head;
There once were brooks sweet whimpering down
    the vale:
The brook's no more – kingcup and daisy fled;
Their last fall'n tree the naked moors bewail,
And scarce a bush is left to tell the mournful tale.

In Clare's poetry the loss of intimately known places is inextricably bound up with the loss of boyhood 'joys' and innocence. In 'Childhood' (which begins 'The past it is a majic word / Too beautiful to last') he describes the games he played when he was a boy, how he and his friends made posies of weeds, and planted the rootless flowers out as make-believe gardens. They had weed picnics, too:

The mallow-seed became a cheese
The henbanes loaves of bread
A burdock leaf our table cloth
On a table stone was spread
The bindweed flower that climbs the hedge
Made us a drinking glass
And there we spread our merry feast
Upon the summer grass

These experiences are what 'time hath stole away' – though in his fierce battle-hymn against enclosure, 'Remembrances', he makes it clear that it is not just the inexorable passage of time that is responsible for his sense of loss, but the greediness of 'Buonaparte' landowners:

By Langley Bush I roam but the bush hath left
    its hill

On Cowper Green I stray, 'tis a desert strange
   and chill
And spreading Lea Close Oak ere decay had
   penned its will
To the axe of the spoiler and self-interest fell a
   prey

Cowper Green today is a huge and featureless arable field. The process of agricultural intensification and social narrowing that began with the parliamentary enclosure of the parish has reached its logical conclusion. A complex ecosystem and community resource has become a mono-culture. In his tribute to the common, written before its destruction, Clare dreams of the people who may have used the place, digging sand, gathering medicinal weeds, or just relishing, like him, the obstinate vigour of the unlovely: 'hemlock with its gloomy hue', stinking hen-bane and furze wreathing 'Its dark prickles o'er the heath'.

And full many a nameless weed,
Neglected, left to run to seed,
Seen but with disgust by those
Who judge a blossom by the nose.
Wildness is my suiting scene,
So I seek thee, Cowper Green!

❧

Not all improving farmers were arrogant Bonapartes. A few looked with real curiosity at the plants that grew on their acres, and wrote about them with a poet's clarity.

In 1748 a young Finn called Pehr Kalm, a disciple of the Swedish naturalist Linnaeus, travelled to England to

study and write about the progress of the agricultural revolution. He came especially to see a celebrated improver,
William Ellis, who farmed at Little Gaddesden in the Chilterns, and who was experimenting with different methods
of weed control and pasture management. Between them
the two men left a record of another aspect of weeds' cultural history: how they might actually be useful inside the
ecology of the farm and the economy of the home.

William Ellis's books on farming are written in forth-
right style, and he's at pains to point out that he is not
so much an innovator as a champion of the best traditional practices, so long as they were tailored to local conditions. In *The Practical Farmer, or Hertfordshire Husbandman*,
for instance, he's rather faint in his praise for Jethro Tull's
new horse-hoe, which 'was a pretty, ingenious Contriv-
ance to save the Expense of Men Hoeing (which generally is 7s. an Acre in all)', provided the right plough was
used. The 'common Wheel plough of Hertfordshire' was
no good because it 'kept the Share-point from coming
near enough the Rows of Beans to turn up the Mould on
their Roots, and so kill the Weeds; so that they were in
a great measure choaked at Harvest'. The more intimate
'Vale Foot-Plough' was the thing to use, since you could
set its shares closer to the crop-rows. But he preferred the
hoeing of weeds by hand, despite the cost.

The most pernicious weed – and most troublesome to
the weeders – was a periodic invader of the pea fields he
calls 'Langley-Beef'. It sounds magnificent, a stout weed
of Olde England. In fact, the name is a vulgarisation of
*langue du bœuf*, the French term for the brash and bristly oxtongue, made intelligibly local – and pronounceable – by
incorporating the name of a village (Kings Langley) five

miles east of Gaddesden. The name occurs nowhere else in Britain, as far as I have been able to find out, though John Gerard, for instance, uses the almost-French 'Lang-de-Beefe', and explains that the plant is so-called because its leaves resemble an ox's tongue. They feel rough and pimply, from the swollen papules at the base of what is a kind of vegetable stubble, and the whole plant has the jizz of a street hooligan. (My friend Mark Cocker called it a 'thug' the first time he saw it.) Oxtongue still appears occasionally on the edges of wheatfields around Little Gaddesden, and in Ellis's day was a real problem. 'This I cannot say will utterly destroy a Pea-Crop, but it will so cripple it, as not to be a quarter Value. It comes up thick, and blows like Sow-Thistle, that when the Peas are mowed, or hooked, the Weed generally disturbs the Workmen with its Flew, or Down, that they are forced to drink much; and what is very particular, this Weed comes perhaps but once in a Man's Life, and sometimes often, so that the Farmers are at a loss to account for it: but it is remarkable, that it never hurts the Bean.' As for Clare, there could be no generalisation about the exacting tastes of indigenous wildings.

Ellis also understood about nitrogen fixation by leguminous crops, and that clover especially could also be an effective and respectful way of controlling weeds – a lesson which is only now being rediscovered as weeds become resistant to chemical herbicides: 'Clover-Grass . . . also affords a sort of dressing to the Ground after 'tis ploughed up, and above all saves that great expense which many have been yearly at for weeding their Ground; which is by this Grass entirely got, and also the damage prevented that the Corn generally sustains by the Weeders treading amongst it: so that it may be depended on, nothing better clears the Ground of trumpery and weeds than a good Crop of Clover.'

Pehr Kalm visited Ellis's farm during the last week of March, before any of the wild plants of the pastures and fields were in flower. So he was obliged to use an indirect way of establishing the mix of species in what were obviously very productive grasslands. He sorted through the dried hay in the barns (a technique which is still sometimes used by field ecologists), and identified twenty-four species, of which only nine were grasses. The remainder were broadleaved plants, including several that would be regarded as grassland weeds today – hoary plantain, daisy, yarrow, knapweed, hawkweed. Remarkably, the plant that 'absolutely predominated' in the hay was the familiar lawn weed, bird's-foot-trefoil, John Clare's 'lambtoe', otherwise known as Lady's fingers. Kalm takes a specimen back to Ellis, and has him confirm that 'this was the Lady finger grass which he praised as beyond compare and set before all other grass species in his *Modern Husbandman* . . . to be in the highest perfection the most proper hay for feeding saddle-horses, deer, sheep and rabbits, as well as cattle – and further exalted phrases'. (Modern research has shown that many of these now-despised weeds of grassland have much higher nutritional value than the fodder grasses amongst which they attempt to grow but are usually sprayed out. The mineral cobalt, essential for nutrition in ruminants, reaches concentrations in plantains and buttercups 160 times that in grasses. Dandelion, stinging nettle and thistles have up to five times the proportion of copper as grasses, and the same species have one and a half times that of iron. Magnesium, a deficiency of which causes 'grass tetany' in grazing animals, has concentrations in grasses of about 0.4 per cent, but over 1 per cent in chicory, ribwort plantain and yarrow.)

Kalm stayed on into April, examining other ingenious and frugal local practices. Snails, very injurious to arable crops in this chalky area, were fed to pigs, which became so fat their bristles fell out, and 'yielded the most delicate and tasty meat that could be obtained'. Holly bushes were clipped so that washing could be hung out to dry on them. Bunches of wild thyme and dog-roses were set into cracks in the walls of the nearby stone mines at Totternhoe, where, seemingly relishing the humidity, 'they would remain green, fresh and sweet-smelling for a couple of months'.

He also gives a minutely detailed account of the use of gorse, or furze, as a fuel plant, especially for bread ovens. Furze grows abundantly on the acid plateaus of the Chilterns, and is conventionally regarded as a detested weed when it invades rough pasture, despite being a highly nutritious food for grazing animals. But it was tolerated on the commons because it was valuable as a fuel. Kalm's descriptions have an almost Clare-like exactitude:

> Because it was continually being cut down by people for fuel, it was now little higher than a hand's breadth. A couple of boys went out to one place, and with a kind of scythe cut it down close to the ground . . . the thickness of the blade was about ¼ inch. It was only sharp on one side, so that it could only be used by a right-handed man, or by holding the handle in both hands, with the right hand nearest to the blade. The iron blade of the scythe fitted into the shaft, which was made of wood, and

that part of it which was made of iron and which fastened into the shaft, had a very sharp angle. So that the man who used it need not stoop while he was cutting with it . . . With this the boys cut down the furze, bracken, old grass and whatever else was needed, raked them together into heaps, and bound them into bundles. They used the thin briars of bramble bushes as twine to tie them up. It was very necessary for those who tied up these bundles to have good mittens or gloves, for both the gorse and the blackberry are among the thorniest kinds of bushes.

The local bracken was harvested for an even wider range of uses. It was cut and stacked in ricks, and used for

all sorts of purposes as a fuel in the place of wood . . . During my perambulations around and about Little Gaddesden, I always found it growing profusely on the common grazing lands and hills . . . We saw people cutting it and collecting it for fuel in a number of places. In the Duke of Bridgewater's park, which lay close beside Little Gaddesden, there was a large brick yard, where large numbers of bricks were prepared. The fuel that was usually put into the kilns to burn the bricks consisted of small bundles of beech twigs and more particularly of this bracken. We saw large heaps of it thatched with straw and lying in the brick yard. People said that on burning, this bracken gives off a much more intense heat than many kinds of wood . . . A local worthy told me that from long experience he could testify that bracken is reckoned among the best of fuels. He used it for baking bread and for much else. In many places it could be seen that bracken was collected, mixed with other straw and used as litter for animals in the farmyard, where it rotted and became manure. It was also used on the ground beneath wheat, pea and corn stacks.

For much of my life I lived just a few miles from Little Gaddesden. The local commons still have their sweeps of gorse and bracken. Pehr Kalm's mentor Linnaeus visited them one spring in the 1730s, and reputedly fell on his knees to thank God when he saw the furze in full flower. In 1866 a radical local landowner, Augustus Smith, organised a campaign of direct action that successfully defeated an attempt to enclose a large area of the waste. On the day the fences were torn down, the local people flocked up to Berkhamsted Common and picked token sprigs of gorse to celebrate that the place was theirs again. Until the commons were finally sold off in the 1920s, the locals adhered to courteous and frugal codes to ensure the survival of their weed resource. There was a close-season for the gorse and bracken, between 1 June and 1 September. On 31 August the commoners would listen for the chimes of the parish church at midnight, and go up to stake out their claims, like gold prospectors.

# 7

# *Gallant-soldier*

## The weed as mercenary

~

OVER THE YEARS I've been lucky enough to work on several occasions with the Royal Botanic Gardens at Kew, the central nervous system of global botany. In the 1980s I wrote a television documentary about the history of the Gardens, which focused on the Victorians' fascination with 'the Wonders of Creation', and on how the ceaseless parade of exotic new plants that explorers were bringing home from the Empire was seen as kind of divine blessing on the nation. Some years later I went to report on how the 1989 hurricane had reduced much of that collection of vegetable marvels to a tangled wreckage. Life, needless to say, went on. Days after the great storm, I watched Kew's scientists crawling amongst the upturned roots of their specimen trees, transfixed by the networks of symbiotic fungal roots that had suddenly become visible.

Between those two insightful moments I spent one summer month working on a book about Kew's prestigious collection of plant illustrations. The pictures are stored in the Herbarium, and filed by species (not by artist) amongst the similarly ordered dried plants. Exquisite eighteenth-century paintings of roses are sometimes the thickness of a sheet of cartridge paper away from the desiccated petals of the original: two different kinds of representation of the living plant.

It is an extraordinary collection, of more than a million pieces. There are lavish paintings by masters such as Redouté and Ehret, drawings by young Scottish draughtsmen recruited for expeditions to the colonies, stylised pictures by Mughal painters working to the East India Company's instructions. There are meticulous diagrams by microscopists, and impressionist watercolours by retired embassy officials with time on their hands.

While I was browsing through this immense visual record of Kew Gardens' global reach, a party of mineralogists came to examine some of the Herbarium specimens from West Africa. They were interested in a group of weedy species which had the ability to extract metals from the soil and deposit them in their leaves. (Spring sandwort does this on old lead-mine spoil in northern Britain.) Traces of metal remain in the leaves even in long-dried specimens, and are detectable by chemical analysis. The leaf becomes a kind of litmus paper. A high mineral content suggests that the soil in which the plant originally grew might be rich in metallic ore. And there was another factor the mineralogists were interested in. All specimens in the Herbarium are labelled with the exact location at which the plant was gathered. They are, in effect, ready-made signposts for exploratory digging.

I suppose I was mildly shocked at this commercial intrusion. I certainly found it ironic that weeds – of little economic value in themselves – should have joined the great roll call of contributors to the exploitation of the planet since the eighteenth century.

~

Plants of all kinds played a central role in Europe's imperial expansion. They were the shock troops through
which the colonial powers imposed their own economic
priorities on foreign cultures. In the eighteenth and nineteenth centuries the core of what the historian Alfred W.
Crosby has called 'ecological imperialism' was the con-
version of traditional self-sufficient agricultures to plantations of profitable non-native plants for export – rubber,
breadfruit, opium, sisal – and later of exotic species for
the greenhouse and herbaceous border. The botanic gardens, and Kew especially, served as coordinating centres
for these projects, disseminating scientific information,
investigating cultivation techniques, selecting promising
cultivars, and importing and exporting the plants themselves. Kew's role in spreading the cinchona tree across
the globe from its native South America, for instance, was
crucial. Cinchona bark is the source of quinine, the only
effective treatment for malaria in the nineteenth century,
and without readily available local sources of this drug,
the European colonisation of Africa and India would have
been stopped in its tracks by disease.

At a more trivial level, Kew was involved in the populari-
sation of what was to become one of the nation's favourite
shrubs, the rhododendron. In the 1830s Kew went into a
temporary decline, and in 1841 the government appointed
Sir William Jackson Hooker as director, with the explicit
recommendation that Kew should take on the responsibility of linking science, public interest and colonial expansion. 'A national garden', the report suggested, 'ought to
be the centre round which all minor establishments of the

same nature should be arranged . . . receiving their supplies and aiding the Mother Country in everything that is useful in the vegetable kingdom. Medicine, commerce, agriculture, horticulture and many valuable branches of manufacture would benefit . . .' Seven years later William Hooker despatched his son Joseph to explore and collect plants in the eastern Himalayas. He discovered, and brought back the seeds of, twenty-eight new species of rhododendron. They were a sensation with the gardening public, perfectly satisfying the growing taste for informal shrubs. No one could have anticipated that some of them would escape to become one of the most invasive weeds of Britain's western woodlands.

Weeds, of course, were rarely spread about deliberately. But, ever the opportunists, they took advantage of this unprecedented movement of plant material across the globe, shuffling around on its coat-tails (or perhaps, with Edward Salisbury in mind, in its turn-ups). Britain was the landfall, deliberate and accidental, for huge numbers of foreign plants, and our weed species probably doubled in number during the eighteenth and nineteenth centuries. Some simply rode piggy-back on crop and garden plants, their seeds cosseted in root-balls or stuck to containers, but always insidiously *there*. Others were welcomed as valuable food plants or glamorous ornaments, but escaped or were thrown out and *became* weeds as a consequence of unforeseen bad behaviour.

One of the classic stories of undercover plant infiltration, a gothic tale of city back-streets and smoky railway carriages, has a botanic garden in a lead role. Oxford University's Botanic Garden was founded in 1621, and is the only such institution to be remembered in the official English name of a plant species. The rather scruffy yellow-flowered daisy that would become known as Oxford ragwort was reputedly first spotted there by Sir Joseph Banks – a botanist explorer who had sailed with Captain Cook – in the mid-eighteenth century, and was officially recorded by the university's Sherardian Professor of Botany, John Sibthorp, in 1794. How it arrived isn't known. It grows naturally on the volcanic clinker of Mount Etna in Sicily, and may have been brought back to Oxford from a botanical Grand Tour. Equally its seeds may have stowed away in the roots of some more obviously attractive Mediterranean species, and ramped spontaneously about the garden stonework. But Sibthorp, following scientific protocol, gives a Latin description of its habits that in Latin reads like a hushed account of an ancient Roman fresco: 'Sub ipsis denique muris urbis rariores stirpes oculis occurrunt, quae tamen cum peregrina sint facie, dubito utrum inter indigenas enumerandae sint . . . quae late se propagans undequaque prorepit, et tapetis instar circa rudera et antiquiores muros sternitur.' (Roughly: 'Finally under the very walls of the city rarer plants meet the eyes, which however, since they are of exotic appearance, I doubt [whether they] should be reckoned among the natives . . . which, propagating itself widely, creeps out from every quarter and is spread like a carpet around the rubble and older walls.')

Within a few years the ragwort had escaped from the garden (which is sited opposite Magdalen College) and begun its westward progress along Oxford's ancient walls. Its downy seeds seemed to find an analogue of the volcanic rocks of its original home in the cracked stonework.

It leap-frogged from Merton College to Corpus Christi and the august parapets of Christ Church, then wound its way through the narrow alleys of St Aldate's. It got to Folly Bridge over the Isis, and then to the site of the old workhouse in Jericho, where, as if recognising that this was a place of poverty, threw up a strange diminutive variant, a type with flowerheads half the normal size (var. *parviflorus*). Sometime in the 1830s it arrived at Oxford Railway Station, the portal to a nationwide, interlinked network of Etna-like stone chips and clinker. Once it was on the railway companies' permanent ways there was no holding it. The seeds were wafted on by the slipstream of the trains, and occasionally travelled *in* the carriages. The botanist George Claridge Druce described a trip he took with some on a summer's afternoon in the 1920s. They floated in through his carriage window at Oxford and 'remained suspended in the air till they found an exit at Tilehurst', twenty miles down the line. The poet Geoffrey Grigson added an atmospheric 1950s gloss on these bizarre commutings: 'between Swindon and London you may see them often enough during the summer, the sunlight through the carriage window catching them as they float about in the thin smoke of cigarettes'.

The plant had reached London by 1867 and Swindon by 1890. In 1899 it was, according to Sowerby's *English Botany*, also on old walls and waste ground at Bideford in Devon and on 'Allersey Church, Warwickshire'. A skim of turn-of-the-century county floras shows that it had also reached sites in Suffolk, Kent, Somerset and Herefordshire. By 1915 it was as far north and west as the Clyde and Caernarvon. But it was the Second World War – when so many city centres were firebombed to the condition of volcanic rubble – that gave Oxford ragwort its second wind. In the 1940s it was the third most widespread weed in Edward Salisbury's survey of London bomb sites, and by 1944, with typically weedy promiscuity, had hybridised with its close cousin, sticky groundsel, to create what became known as London ragwort.

Kew Gardens, like Oxford, also had an errant ragamuffin that went on to colonise most of Britain. In 1793 the Gardens received a specimen of the diminutive Peruvian daisy *Galinsoga parviflora*, named after the splendidly ennobled Spanish botanist Don Mariano Martinez de Galinsoga. The plant itself is less grand, with small, dingy white flowers on a lax stem. In the 1860s it escaped from the Gardens and became locally established in gutters and pavement cracks. For a while it was known as Kew Weed. But once its airborne seeds had blown further afield, and into less salubrious neighbourhoods, it needed a more general and more down-to-earth tag. *Galinsoga* was too much of a mouthful for south Londoners, so they vulgarised it to gallant-soldier – a name that has stuck, I suspect, partly because it is so ironically inappropriate. (In Malawi, where this weed of very unmilitary bearing is also naturalised, it is known as 'Mwamuna aligone' – 'my husband is sleeping'.)

And so they poured in, not just via the botanic gardens, but in imports of agricultural seed, the soil of fashionable pot-plants, the raw materials of brewers and wool merchants. Thanet cress, now known as hoary cress, reputedly arrived in Britain in the wake of a battle on the Dutch island of Walcheren during the Napoleonic wars. The casualties were brought to Ramsgate on palliasses stuffed with hay, which contained seeds of this north European weed. The hay was later disposed of to a local farmer, who ploughed it into his fields. The cress seeds sprouted, colonised the Thanet area, spread along the whole south coast, and then advanced across most of southern England. The now pan-global weed Canadian fleabane arrived in Europe in the seventeenth century in the stuffing of a bird imported from North America. Pirri-pirri-bur came from the Pacific on imported fleeces. It's a low mat-forming perennial from the open country of Australia and New Zealand, and has curious globular flowerheads, adorned with spines, like a miniature mace. They become entangled with the animals' fleeces and can end up anywhere that wool-trade waste (known as 'shoddy') is used as fertiliser. Pirri-pirri-bur is now thoroughly naturalised on some sand dunes in East Anglia and southern England, where its seeds stick to the clothes of romping children as obstinately as they do to grazing animals. Weeds have also been spread about via the sweepings from pet-food shops, the dumping of ballast from ships, and from discarded Asian takeaways. The spread of the fruitily perfumed pineapple weed, which arrived in Britain from Oregon in 1871, exactly tracked the adoption of the treaded motor tyre, to which its ribbed seeds clung as if they were the soles of climbing boots.

The commonwealth of the arts has been as effective a conduit as commerce for new weeds. Ivy-leaved toadflax's miniature blue and yellow snapdragons now ornament old walls in almost every parish in Britain. It is probably only regarded as a weed by municipal fusspots, worried about the tidiness of their stonework – except that it is an alien plant very much in the wrong place. It's a native of mountains in southern Europe, and didn't arrive in Britain until the early seventeenth century. The story is that its seeds were caught up in the packing of some marble statuary imported from Italy to Oxford, whence, like the city's eponymous ragwort, they migrated into the wider world via the college walls. (It was called 'Oxford weed' for a while.) John Ruskin adored ivy-leaved toadflax for its graciousness and Classical echoes. Visiting the church of the Madonna dell'Orto in Venice in 1876, he found that the exquisitely painted flower growing by the side of St Peter in Cima da Conegliano's portrait of a group of saints was the same *erba della Madonna* that grew on the marble steps outside. In his diary for 16 September he writes: 'I am weary, this morning, with vainly trying to draw the Madonna-herb clustered on the capitals of St Mark's porch; and mingling its fresh life with the marble acanthus leaves which saw Barbarossa receive the foot of the Prince of Christendom on his neck.' Henceforth the toadflax became his signature plant, a symbol of the detailing of nature.

Its popular English names are more secular. Travelling Sailor and Mother-of-thousands reflect one of the plant's weedy characteristics, its capacity to rapidly invade niches which resemble its original habitat. It has an intriguing mechanism that enables it to colonise walls upwards. When it's in bloom (throughout the year in England's currently mild climate) the flower-stalks bend towards the light; when the flowers are finished the seedheads bend the other way, so that the seeds are more likely to be shed into cracks and mortar joints in the supporting stones. (Its anatomy has fascinated children. The folklorist Ray Vickery recorded this exchange with a girl in Dorset in

1983: 'This [ivy-leaved toadflax] is what we call wall rabbits.' 'Why do you give it that name?' 'Because if you turn the flower upside down, and squeeze its sides, like this, it looks like a rabbit's head.')

A similar exodus of weeds from Italy happened via the works of the Danish sculptor Bertel Thorwaldsen. When he died in Rome in 1844 Thorwaldsen's statuary was brought home to Copenhagen. When the well-cushioned cases were unpacked, a sprinkle of seeds fell out of the straw, and the following year twenty-five Italian species, many of them Mediterranean weeds, had sprung up around the premises. A few became naturalised in Copenhagen, and some were preserved and specially cultivated in a garden dedicated to Thorwaldsen's memory.

The modern equivalent to the train-hopping of Oxford ragwort is the trunk-road hitchhiking of Danish scurvy-grass. Up to the 1980s *Cochlearia danica* was a scarce native of the drier areas of the coasts around Britain. It grew (and still does) on clifftops and sea walls and the landward side of salt marshes. It resembles common scurvy-grass in its fleshy leaves and small four-petalled white flowers, but must have some latent weedy gene that its commoner cousin lacks, some intrinsic willingness to tolerate facsimiles of the true shoreline. In the mid-1980s it began to appear on a few inland railway-line sites, where its seeds had been introduced with stone rubble brought from the seashore. Then it began to show up along the edges of motorways and major roads. The plants were packed close together, especially on the central reservations, and in their flowering time of March and April it was as if a deep and persistent frost had gripped the verges.

By 1993 there were patches on stretches of major road in over 300 locations. In 1996 I did a rough paper survey of the nationwide distribution. It read like a commercial traveller's route-map. There were concentrations on the M4, M5 (especially close to Cheltenham and Cardiff), the M6 and M56; along many reaches of the A1, the A5 in Anglesey, A11 in Suffolk and A30 in Devon. It had crossed the Scottish border and appeared alongside the A74 in Dumfriesshire.

But there were almost no roadside records for Ireland, despite Danish scurvy-grass's native presence on Irish shores. What is unusual about the road system in Ireland is that, unlike Britain's, pure grit, not salt, is used to treat its roads during winter freeze-ups. There are doubtless many factors that have enabled *C. danica* to spread inland: the turbulence of articulated lorries whirling its seeds along, for instance, and the similarity between shingle strandlines and the bare, stony edges of major roads. But the saltiness of the modern road – that shoreline tang sprayed from council gritting lorries every icy evening, even in the landlocked heart of Britain – has been the crucial factor. Again, a social innovation has been immediately exploited by a weed.

Danish scurvy-grass continues its viatical invasion. It has penetrated (along with the Eurolorries) far beyond the trunk-road system – but not beyond the salt. A couple of hundred yards from our house in Norfolk an ungraded road runs north–south. A couple of years ago the first clump of Danish scurvy-grass appeared at one end of it. It's at the junction between the lane and main road to Thetford, down close to the tarmac and below the province of the celandines and cowslips. But in the severe winter of 2009–10 the lane itself was salted and the scurvy-grass has crept a few feet north. In a few years it may rime the whole laneside, a last sparkle of winter white before the spring yellows begin to glint through. It needs a friendlier and less awkward name than Danish scurvy-grass. Wayfrost might do.

For all their invasions of ancient walls and motorway verges, only a few of Britain's immigrant weeds have become troublesome in the working world. Most, though they may have become more widespread, have remained in the margins. Oxford ragwort haunts car parks and railway lines but has been almost exterminated on the walls of Oxford. Gallant-soldier rarely jumps from the street into the garden. And Canadian fleabane, though it abounds in waste places across Europe, hasn't yet evolved into an agricultural pest. But weeds have moved out of the Old World as well as into it, and met with a very different kind of reaction.

∼

In New York's Central Park there is a garden devoted to growing all the plants mentioned in Shakespeare's works. It was inaugurated in 1916, on the 300th anniversary of the playwright's death, a follow-up to a more risky nineteenth-century project to introduce all Shakespeare's birds to the United States. But plants (or at least their seeds) also have wings. Titania's wild thyme and oxlip have doubt-less shown proper decorum, and stayed exactly where they were planted. But Lear's 'hardocks, hemlock, nettles' jumped the wall at the first opportunity – which might

have been a cause for concern if these European species
had not already found their own way to America centuries
earlier.

Just as Mediterranean farmers inadvertently took
their weeds with them when they migrated north and
west across the Continent, so the next wave of European
expansion, across the Atlantic, carried the weed diaspora
into the Americas. The temperate regions of North and
South America (and of Australia and New Zealand in
the southern hemisphere) were attractive destinations
for European emigrants. They had similar climates to
Europe, few dangerous diseases or large predatory animals, and could sustain European crops. They also suited
European weeds, which had made the crossing caught up
in the settlers' clothes and seed corn and animals' hooves.
As early as the sixteenth century Spanish writers were
reporting outbreaks of *malas hierbas* in the grasslands of
Mexico. They make a familiar list: thistles, plantains, nettles, nightshades, docks, wild oats. There were also, more
pleasantly for the settlers, forage plants such as clover
and meadow-grass, which had quite likely arrived on, or
in, their livestock. White clover was so widespread in the
pampas that even as early as 1555 the Aztecs had given it a
name: *Castillan ocoxchitl*.

The weeds moved north with the Spanish, and new
waves arrived with the seventeenth-century English settlers. John Josselyn, who visited New England in 1638 and
1663, made a list 'Of Such Plants as have sprung up since
the English planted and kept Cattle in New-England'. It's
worth reproducing in full, because it is a refrain that goes
back to the European Neolithic:

Couch grass
Dandelion
Sow-thistle
Night Shade, with the white flower [black
    nightshade]
Nettles, stinging
Plantain
Wormwood
Patience [patience dock]
Adder's tongue
Cheek-weed [chickweed]
Compherie, with the white flower
The great clot-bur [burdock]
Shepherd's purse
Groundsel
Wild arrach [probably fat-hen]
Mallowes
Black henbane
Sharp-pointed dock
Bloodwort [wood dock]
Knot-grass
May-weed
Mullin, with the white flower

Greater – or 'ratstail' – plantain had by this time been nicknamed 'Englishman's foot' by the Native Americans, who had witnessed its prodigious advance in the white man's wake.

It was the success of English grasses that had the most profound effect. A few species had been brought over deliberately. William Penn records sowing a mixture in his courtyard as early as 1685. But the seeds of couch and foxtail and meadow-grass would have arrived anyway, caught up in the tails and hooves of the settlers' animals, and they found the New World a *tabula rasa*, ripe for colonisation. Along the eastern seaboard, the settlers were cutting down much of the native forest. In 1629 Captain John Smith reported that most of the woods around Jamestown, Virginia, had been 'converted into pasture and gardens; wherein doth grow all manner of herbs and roots as we have in England in abundance and as good grass as can be'. The new, light-loving grasses had few competitors in areas that had been under permanent tree cover. And they had one crucial advantage over the indigenous grass species. Over thousands of years they had evolved to cope with the pressure of voracious grazing animals. The more they were chomped by domestic sheep and cattle the more adept they became at developing underground roots, spreading sideways, forming mats, regenerating through multiple shoots. By the time they were introduced to America, European grasses had learned to actively thrive under grazing. Thomas Budd, writing in Pennsylvania in the late seventeenth century, described the insidious takeover exactly: 'If we sprinkle but a little English hay-seed on the Land without plowing, and then feed sheep on it, in a little time it will so encrease, that it will cover the land with English grass.'

But it was a different matter for North American grass species. East of the Mississippi the native grasses had never experienced anything like the immense herds of bison that roamed the Great Plains. They were grazed by deer, but these were lighter and more selective in their feeding, and had nothing like the effect of a concentrated body of dairy cows or determined goats. So they were simply overwhelmed, munched into oblivion, and the resilient European grasses moved in, helped by the way the ground was churned into a muddy seed-bed by the animals' hooves.

The new grasses followed the settlers, north, south and west, occasionally getting ahead of them in places where traders or advance parties had briefly rested. One Illinois pioneer noted in his journal: 'Where the little caravans have encamped as they crossed the prairies, and have given their cattle hay made of these perennial grasses, there remains ever after a spot of green turf for the instruction and encouragement of future improvers' – though sometimes the spot of green turf had, in high summer, the slightest sheen of blue. When the southbound wave of settlers crossed the Appalachians and reached Kentucky they turned one English weed not just into a dominant forage-plant, but a symbol of the rural south. Smooth meadow-grass, *Poa pratensis*, is a common, widespread but unexceptional species of grassy places in Europe. It grows about two feet tall, and has an adventurous root system. When it is in flower, the heads give the impression of a slight blue haze. In the small clumps in which meadow-grass normally grows in Europe this toning isn't always obvious. But in the uncontested new grazing lands of North America it could colour whole sweeps of grassland. It's not clear exactly when this subfusc European weed was honoured with the name Kentucky blue-grass, but the tag stuck. It seemed to capture something about the local *zeitgeist*. Southern fiddle-music picked up the name of 'Blue-grass' and has since become one of the most distinctive brands of Country and Western music. And in 1936 Florence Graham, founder of Elizabeth Arden, named what would become one of the world's classic perfumes 'Blue Grass' to 'recall the view from her home in Virginia'.

It was, for a while at least, different on the other side of the Mississippi. The native buffalo and grama grasses, which had evolved in the presence of huge herds of wild grazing animals, were able to tolerate European domestic cattle and didn't succumb to the parallel advance of European weeds. Only where the ancient swards of the Great Plains were ploughed up for wheat were the invaders able to get a roothold, and this process escalated after the wholesale slaughter of the buffalo herds in the nineteenth century.

By the 1860s invasive European weeds had reached the far west, and for once, were welcomed. The Gold Rush of 1849 had generated a huge demand for beef, which resulted in over-grazing of the Californian grasslands. This was followed by floods in 1862, and then a two-year drought. When the rains finally came, it was the European invaders that sprouted first and grew most densely, and prevented vast areas of good agricultural land from having their topsoil washed away. At this time there were reckoned to be more than ninety alien weed species established in the state. By the twentieth century two-thirds of the vegetation of the western grasslands was composed of introduced species, mostly European. By the end of the century, of the 500 most significant agricultural weeds in North America, 258 were from the Old World.

The curious thing is that this has been a very one-sided invasion. Although large numbers of American plants have become naturalised in waste places in Britain (Michaelmas daisy and Canadian fleabane are conspicuous examples), not a single North American species has become a troublesome weed of farmland, and few could even be described as invasive. This imbalance mystified nineteenth-century

American naturalists, and Charles Darwin, in an untypically frivolous and incurious mood, couldn't resist making a joke about it at the expense of his botanist friend Asa Gray. 'Does it not hurt your Yankee pride,' he wrote in a letter, 'that we thrash you so confoundedly? I am sure Mrs Gray will stick up for your weeds. Ask her whether they are not more honest, downright good sort of weeds.' She responded with an answer that was both witty and scientifically spot-on. American weeds, she wrote, were 'modest, woodland, retiring things; and no match for the intrusive, pretentious, self-asserting foreigners'.

Alfred W. Crosby, who gathered so much of this story together in his book *Ecological Imperialism*, paints a vivid picture of the general conditions in which weeds become dominant.

> What does 'Europeanized' mean in this context? It refers to a condition of continual disruption: of plowed fields, razed forests, overgrazed pastures, and burned prairies, of deserted villages and expanding cities, of humans, animals, plants and microlife that have evolved separately suddenly coming into intimate contact. It refers to an ephemeralized world in which weed species of all phyla prosper and other life forms are to be found in large numbers only in accidental enclaves or special parks.

# 8

# *Burdock*

## 'Leaves wherewith to adorn foregrounds'

DESPITE THE TROUBLE they cause, weeds have always had apologists seeking to explain their existence on the earth and find some moral teaching in their lifestyle. For the eighteenth-century school of 'physico-theology' (a prototype of the modern theory of Intelligent Design), for example, they had two kinds of usefulness. First, as demonstrations of God's canniness as a botanical engineer; second, as salutary scourges of human arrogance. Painters, too, found in some weeds a kind of epitome of natural dignity. From the mid-seventeenth century, Shakespeare's despised 'hardocks', the expansive, floppy-leaved, adhesive-fruited burdock, began to feature in landscape paintings. It's never centre stage, nor obviously significant. But it lurks in the margins of a multitude of pictures – felted, foppish, sometimes hard to make out, as if it were some kind of emblem whose meaning the viewer had to decipher. It was the first weed to be credited with some kind of artistic – or architectural – beauty.

Burdock is one of the least likely weeds to have found itself in this exalted position. It's a tall, rank and, to many eyes, gawky species (two species, in fact – greater burdock, *Arctium lappa*, and the commoner lesser burdock, *A. minus*, both very variable and often confused). In both species the thick stem has large heart-shaped leaves at the base and branching stems carrying smaller leaves higher up, culminating in spires of thistle-like purple flowers that turn at seed-time into bristly spheres – the burrs. It haunts woodland clearings (probably its native home), roadsides, field edges and the waste patches round gardens and derelict buildings. One later artist remarked that the leaves 'have a messy droopingness – they seem to be crawling along the ground'.

Images of burdock first crop up in seventeenth-century Dutch painting, indistinct in the corners of a few landscapes by Jan Wynants and Jacob van Ruisdael. In the work of Claude Lorrain, widely regarded as the father of European landscape painting, it becomes more obvious. A modest tuft, its leaves mantling the rocks, sits in the bottom right-hand corner of *Landscape with Dancing Figures* (1648). Behind it, young people picnic and jig with tambourines. In the more wistfully shaded *Landscape with Rustic Dance* (1640–41) the grey-green fronds have moved to the bottom left-hand corner. In Claude's best-known painting, *Landscape with Narcissus and Echo* (1645), they are still at the bottom of the painting, but more central, and the arch of the leaves echoes Narcissus's splayed legs and arms as he gazes down at his reflection in the water. In *Landscape with David and the Three Heroes* (1658), which features a lot of men with spears, the burdock (still perched on the bottom edge of the scene) is at last allowed to show a flowering spike.

In these and many other pictures the burdock is not there just as token greenery. There are plenty of flowers and token foliage round the borders of Claude's pictures, but most of them (with the exception of the wild daffodil in *Narcissus and Echo*) are purely stylistic – notional, daisy-like discs that recall the floral decoration of the medieval books of hours. Burdock is the only one that is drawn with realism, and is instantly recognisable.

Thomas Gainsborough borrowed much from Claude, including burdock, and a token tuft occupies a typically Claudean position in the bottom right-hand corner of *The Cottage Door* (1780). It acts as an ornamental base for the dead and gloomy tree trunk which frames the mother and children in the cottage doorway. In his famous Suffolk study *Cornard Wood* (1746–7) the felted leaves, looking as flat as huge toadstools, tumble down a mound below an oak tree – exactly where you might find them in a wood. He also made a small, intimate sketch of a tuft of burdock (late 1740s) which begins to hint at why the plant held such a fascination. The leaves, outlined with a few bold strokes in black charcoal, are set against a gnarled tree trunk. They are shown leaning towards the viewer like open hands, palms forward, left and right. Gainsborough catches perfectly their sculptural qualities, the heavy central rib, the wavy, scalloped, almost rococo edges. What burdock suggests in these pieces is that beauty can reside in the uneven and the asymmetrical – in the idea of weediness, in fact.

Almost contemporary with Gainsborough, Joseph Wright of Derby's outdoor portrait *Sir Brooke Boothby* (1781) has his reclining subject's feet comfortably resting in a shoal of burdock leaves near some silver birches, as if they were a kind of nest.

Close inspection will reveal burdock clumps in Richard Wilson, J. M. W. Turner, John Linnell, James Ward,

John Constable and Edwin Landseer. Burdock spotting can become as compulsive as searching for goldfinches in Italian Renaissance paintings, except that, unlike the bird, burdock has no specific symbolism, which is maybe why the Pre-Raphaelites largely ignored it.

The true master of burdock display is George Stubbs, and in several of his pictures the plant is much more than a tonal ornament or filler of awkward spaces. In *A Lion Devouring a Horse* (1769) it plays such an active role in the dynamics of the picture that it ought, by rights, to be mentioned alongside the other organisms in the title. The white horse, muscles so tense they have the look of a dissection diagram, turns its head back in agony towards the lion, which is clinging to its back and tearing at the hide above its ribs. Under the horse's raised right hoof, a burdock leaf shears leftwards, its shape mirroring that of the horse's terror-struck head. But the leaves aren't the blandly smooth grey-green foliage – leaf as carved stone, perhaps – of Claude and Gainsborough. They are picked out in high, mortal detail. Their ribs are as contoured as the horse's. They are beginning to age, wilting at the edges, showing patches of brown rust. One is already dead, a tan husk drooping towards the ground. A weed, Stubbs seems to be suggesting, experiences stress and ageing like any other living thing.

This is an unusual perspective on botanical beauty, that it might take the form of elegance under pressure – what you might call grace.

~

In the world beyond art, burdock's uses and meanings have centred chiefly around its clinging, spherical seeds, the 'burrs', which are equipped with a mass of flexible hooks. They have been used in children's games for centuries (they stick to hair and clothing as readily as to animal fur) and in a bizarre, and still surviving, ritual in the Royal Burgh of Queensferry, Edinburgh. On the second Friday in August a man dressed from head to toe in burrs perambulates about the town, visiting houses and receiving gifts. The parade of 'The Burry Man' is thought to have originated in a fertility rite to bring better luck to the fishing, and there may be echoes of sympathetic magic, with the burrs representing both fish scales and fish hooks.

But burdock's burrs have a more practical modern significance, and one that oddly resonates with the landscape painters' idea of the beauty of irregularity. In the 1950s they became the inspiration for Velcro.

The hook-and-loop fastener is a classic *biological* solution to a problem. It has little in common with the precision of 'hard' engineering. It is, science writer Peter Forbes suggests, 'the first example of fuzzy logic . . . The hook-and-loop fastener doesn't have to be lined up accurately . . . Whether an individual hook goes through a specific eye is irrelevant: every time you use it, enough hooks will find an eye to achieve a bond.' Burdock burrs are a maze of thin spines with flexible hooks at the tip; the 'loops' are the tangle of animal hair they have evolved to snag. The story of how they generated a new material, entirely synthetic but incorporating 'soft' biological technology, revolves around George de Mestral, a Swiss inventor who had a fascination with fasteners. In the 1940s he was in the habit of taking his dog on hunting expeditions in the Jura mountains. When he got home from one walk his dog was covered in burrs, and instead of simply removing them, he began to reflect on their tenacity. (The apocryphal story is that he was frustrated by the difficulties of linking the large hooks and eyes on his wife's dress when they were going out in the evening.) The burr is spherical because it has evolved to maximise the number of angles by which it might stick to passing animals. But de Mestral realised that if it were flat, it might stick to a rough surface at whatever angle they met.

De Mestral had to wait until after the Second World War before he could create his flattened burr. At the time, nylon, invented in 1937, was the only existing material that could be fashioned into the synthetic hooks and loops he required. But nylon was so valuable for the war effort that it was not until the late 1940s that de Mestral was able to obtain enough of it to experiment with. The loops were easy, but the hooks had to be made by passing nylon thread over a bar, which was heated to fix the shape. The original Velcro patent was filed in 1951, and with help from a French weaver de Mestral perfected his fastening system and put it on the market in 1955. (It received the ultimate honour in 1996 of a spoof scientific paper, which, echoing *Panorama*'s famous April Fool's Day 1957 hoax on the spaghetti orchards of Switzerland, commented on the fortunes of the Californian Velcro plantations. They were, at the time, experiencing problems as high winds were causing the spores from the hook bushes to commingle with those from the loops, resulting in internally locked Velcro bolls which were quite impossible to disentangle.)

John Ruskin would have been appalled by this mechanical

exploitation of a wild plant, and by the assumption that the burriness of burdock had evolved to help the plant distribute its seeds. In the first volume of his *Proserpina - Studies of Wayside Flowers* (1874) - he describes how the construction of a burdock's leaf (which he perfectly describes) contributes to its beauty:

> When a leaf be spread wide, like the Burdock, it is supported by a framework of extending ribs like a Gothic roof. The supporting functions of these is geometrical; every one is constructed like the girders of a bridge, or beams of a floor with all manner of science in the distribution of their substance in the section . . . But when the extending space of a leaf is to be enriched with the fullness of folds, and become beautiful in wrinkles, this may be done either by pure undulation as of a liquid current along the leaf edge, or by sharp 'drawing' - or 'gathering' I believe ladies would call it - and stitching of the edges together. And this stitching together, if to be done very strongly, is done round a bit of stick, as a sail is reefed round a mast; and this bit of stick needs be compactly, not geometrically strong; its function is essentially that of starch, - not to hold the leaf up off the ground against gravity; but to stick the edges out, stiffly, in a crimped frill. And in a beautiful work of this kind, which we are meant to study, the stays of the leaf - or stay-bones - are finished off very sharply and exquisitely at the points; and indeed so much so, that they prick our fingers when we touch them; for they are not at all meant to be touched, but admired.

A few pages later, more bluntly, he urges readers to study its structure: 'Take a leaf of burdock – the principal business of that plant being clearly to grow leaves wherewith to adorn foregrounds.'

These are extraordinary and baffling passages, full of intimate glimpses of the engineering of leaves, but seeming to suggest that these exist more for the beatification of the observer than the livelihood of the plant. *Proserpina* is like this throughout. It is a confused and at times deranged attempt to devise a new, anti-Linnaean plant taxonomy, based on aesthetic principles rather than scientific understanding. It passes moral judgements on whole orders of plants, yet sometimes has moments of startlingly original observation and insight, as in this evocation of a poppy flower: 'We usually think of the poppy as a coarse flower; but it is the most transparent and delicate of all the blossoms of the field . . . the poppy is painted *glass*; it never glows so brightly as when the sun shines through it. Wherever it is seen – against the light or with the light – always, it is a flame, and warms the wind like a blown ruby.' This may be the best descriptive passage on the poppy in the English language, and it comes close to offering a poetic intimation of the role of the sun in plant growth, and of the seductive power of the hot scarlet petals to other creatures.

But such plant- or nature-centred views were an abomination to Ruskin. In one of his deeper depressions he remarked with disgust that the theory of photosynthesis made us look on leaves as no more than 'gasometers'. Beauty of form or function in a plant he saw as an abstract quality, planted there by God for the elevation of humans. That it might in some way be 'recognised' by a non-human organism was repugnant to him. That the ruby flame of a poppy, or the intricate anatomy of an orchid bloom, might be attractive – be beautiful, as it were – to an insect was a blasphemy. This led Ruskin to believe in a hierarchy of organisms based on his own aesthetic ideas. 'The perception of beauty,' he wrote, 'and the power of defining physical character, are based on moral instinct, and on the power of defining animal or human character. Nor is it possible to say that one flower is more highly developed, or one animal of a higher order, than another, without the assumption of a divine law of perfection to which the one conforms more than the other.'

Ruskin had in effect devised an aesthetic version of the Doctrine of Signatures. God had 'signed' certain plants with imprimaturs – symmetry of petals, for instance, or the angles between stalk and leaf – which might have some base biological function, but which were principally indices of the divine quality of beauty. It was the responsibility of the cognoscenti to recognise and interpret these signs.

Weeds were plants which fell short of these ideals of perfection, which had mysteriously 'degenerated' from higher forms. At first Ruskin accepts the definition (already current in the nineteenth century) that a weed is a plant in the wrong place – adding, rather parochially, 'whoever saw a nettle or hemlock in a right one?' But he then introduces an interesting, and botanically astute, twist. A weed is 'A vegetable which has an innate disposition to *get* into the wrong place . . . It is not its being venomous, or ugly, but its being impertinent – thrusting itself where it has no business, and hinders other people's business – that makes a weed of it.' So a plant might be innately beautiful and perfect in form, but not if another of its innate qualities was ambition.

But just a few pages later, he begins to outline qualities in weeds which have nothing whatever to do with their impertinent intrusion into other beings' business. 'Hardihood and coarseness of make', for example, far from being seen as marks of experienced resilience, should be viewed as the stigmata of the drifter. 'A plant that can live anywhere, will often live where it is not wanted.' Ruskin's language becomes increasingly haughty, as if the weeds were making moral decisions about how they lived: 'That [a weed] should have no choice of home,' he goes on, 'no love of native land, is ungentle; much more if such discrimination as it has, be immodest, and incline it, seemingly, to open and much traversed places, where it may be continually seen of strangers.' Weeds were vulgar, vagrants who familiarised with the common herd.

They also, he reflected, had contradictory qualities which fitted them for a 'weed's work': 'stubbornness, namely, and flaccidity', which under the influence of some 'serpentine power' can change a pleasing stem into a stake and a leaf into a spine. So – in a chapter largely devoted to the violet – he pours out his invective on every plant with any kind of weedy irregularity, and deplores how the 'the recent phrenzy for the investigation of digestive and reproductive operations in plants may by this time have furnished the microscopic malice of botanists with providentially disgusting reasons, or demonically nasty necessities, for every possible spur, spike, jag, sting, rent, blotch, flaw, freckle, filth, or venom, which can be detected in the construction, or distilled from the dissolution, of vegetable organism'.

Ruskin didn't deny that the forms of plants could be functional. But he passionately denied that they had any significance or value (beyond the purely mechanical) inside the universe of their own lives. A quality like beauty had no connection with the grace and elegance with which a plant lived out its existence on its own terms and amongst its own kind. It could only be granted to them, or withheld, by human beings with the divinely endowed gift of making moral judgements on nature. Which is why he argued that the flower itself was the be-all and end-all of plant existence, not because it was an inspiration to insects and the forerunner of the seed, but because of the pleasure that it gave to *human* eyes.

No one today would pay much attention to this perverse strain in Ruskin's thinking. Much of it was a consequence of his deteriorating mental state and his despair at the advancing ugliness of the industrial age. Yet it distils, in its deranged way, a central theme in human attitudes towards weeds, that they should be judged entirely by our standards, not by those of the organic world they inhabit.

Ruskin's concepts are a world way away from the notion of beauty as *process*, as an expression of *elegance* in the business of living and ageing, as celebrated in George Stubbs's paintings of the leaves of burdock and George de Mestral's insights into the exquisite engineering of its fruits. In 2008 the distinguished American photographer Janet Malcolm took this tradition one stage further and produced a portfolio of twenty-eight extraordinary close-ups of single burdock leaves. She loves burdocks, for their grandeur and individuality. No two are the same. And she admires the way they register their experience – wind-blast, the gnawing of insects – on the expansive canvas of their leaves. In her own notes to the collection she explicitly acknowledges Richard Avedon's photographs of famous people as an inspiration for her portraits of 'uncelebrated leaves'. 'As Avedon sought out faces on which life had left its mark,' she writes, 'so I prefer older, flawed leaves to young, unblemished specimens – leaves to which something has happened.'

So, for three successive summers, she picked burdock leaves, propped them up in small glass bottles, and photographed them head on 'as if they were people facing me'. This is as close as you can come to getting a vegetable to 'put on its best face' for the camera. And the results absolutely vindicate her approach. The burdock leaves are dignified, indomitable, gracious under fire. They are pocked with holes by hailstorms, blotched by strangely beautiful patches of virus and blight, tunnelled by leaf-mining grubs. In the final portrait the leaf has been so comprehensively chewed away by insects that it has been reduced to a skeleton of ribs, a vision of a tree in winter. I think even Ruskin might have admired this image of the elemental structure of a still-surviving leaf.

Of course, many plants show resilience under siege. But to see it so respectfully celebrated in a despised weed is, to me, a revelation of a special kind of wild beauty.

It was the Irish gardener William Robinson who coined the now familiar oxymoron 'the wild garden', and first suggested that the feral beauty of weeds might have a place in our 'outside rooms'. In the late nineteenth century this was a revolutionary idea in gardening. The Victorians had found a perfect way of expressing their twin passions for extravagance and discipline in the institution of 'carpet bedding'. Tender and often garishly coloured flowers from the imperial outposts were raised in heated greenhouses, planted out in straight rows and symmetrical formations (with the exact spaces between each plant kept bare and weed-free for contrast), allowed their brief season of brilliance and then yanked out again. The young Robinson, working as an apprentice gardener on the Ballykilcavan Estate, saw gardeners behaving more like drill sergeants than plant stewards. It was as if, he wrote, they were 'carrying the dead lines of the building into the garden'. Robinson had a vision of a quite different style of planting, in which plants were allowed to mingle with each other as they did in nature, bulbs under trees, ferns invading damp hollows. He thought the regimentation he saw at Ballykilcavan intolerable, and left in 1861, when he was twenty-two years old, amidst rumours that he had stormed out leaving the greenhouse windows open, the stoves out, and carnage amongst the inmates.

He published his classic *The Wild Garden* nine years later, in 1870, after travelling extensively for the Royal Botanic Society's Garden in Regent's Park. His explorations in Europe and America had helped him develop a garden philosophy based on the way plants grew together in the wild. In contrast to Ruskin's abstract and fixed aesthetic, Robinson's was governed by a conviction that naturalness was a *process*, and far from tidy and predictable. He admired the 'mystery and indefiniteness that constitute beauty of vegetation in its highest sense'. The first edition (1870) had a foreword written by the radical writer and wit Sydney Smith (founder of the *Edinburgh Review*) which catches something of the underlying ethos of the book:

I went, for the first time in my life, some years ago, to stay at a very grand and beautiful place in the country, where the grounds are said to be laid out with consummate taste. For the first three or four days I was perfectly enchanted; it seemed something so much better than nature that I really began to wish the earth had been laid out according to the latest principles of improvement . . . In three days time I was tired to death: a thistle, a nettle, a heap of dead bushes – anything that wore the appearance of accident and want of invention – was quite a relief. I used to escape from the made grounds, and walk upon an adjacent goose-common, where cart ruts, gravel pits, bumps, irregularities, coarse ungentlemanlike grass, and all the varieties produced by neglect, were a thousand times more gratifying than the monotony of beauties the result of design, and crowded into narrow confines.

The book's original subtitle was 'The Naturalisation and Natural Grouping of Hardy Exotic Plants with a Chapter on the Garden of British Wild Flowers'. His plan for a wild-flower garden – outlandish at the time – is the first mooting of the idea of growing native plants together purely for their ornamental value. There are plenty of weed species in the list. Celandines by streams, or in damp corners of the lawn; poppy and pheasant's-eye under walls and on stony banks; corncockle, mallow, St John's wort, rosebay willowherb scattered in the beds.

But in a later (1881) edition of *The Wild Garden*, he drops the wild-flower reference. This was only a minor aspect of the book, which was about wildness as a *quality*, not as a species list. Robinson's real mission was to contrive new natural groupings – sometimes compressed or deliberately contrasted – and to combine plants that the accidents of geography had kept apart in the wild. It was a garden, after all, a human space, not a facsimile ecosystem, that he was constructing.

So began a fashion for the deliberate naturalisation of plants from temperate regions across the globe, a wave of potential new weed species that rivalled the accidental introductions of the seventeenth, eighteenth and early nineteenth centuries. The exotic plants that Robinson championed are still in our gardens. They are also way beyond them, and make up one of the most distinctive strains in modern weed flora. Robinson was well aware of the invasive qualities of many of the plants he was recommending. He warned that they be should be allowed to naturalise and spread well away from the formal borders, in shrubberies and woodland at the edges of big estate gardens – where, of course, they were perfectly positioned to jump the garden wall.

He recommended comfrey as ground cover: 'if a root or two of it be planted in any shrubbery, it will soon run about, exterminate the weeds, and prove quite a lesson in wild gardening' – not anticipating that the species would quickly escape and become weeds themselves. White comfrey (a particular favourite of Robinson's, and of mine too) is now the most expansive alien of the road verges near my home in south Norfolk. Goat's-rues are 'tall and graceful perennials . . . among the pretty tall border flowers, and they are useful for planting in rough places', and have become one of the commonest weeds in waste places around London. Japanese knotweeds were barely grown in Britain before Robinson popularised them. He recognised that they 'cannot be put in the garden without fear of overrunning other things, while outside in the pleasure ground or plantation, or by the waterside where there is enough soil, they may be very handsome indeed'. Goldenrods and Michaelmas daisies (both from North America) were his most successful rehabilitations. Species from both families were introduced to Britain towards the end of the seventeenth century, but 'used to overrun the old mixed border and so were abolished'. But Robinson had seen them growing together in New England woods in the autumn, where he thought them 'a picture'. So he recommended trying them instead in 'half-cared-for places in shrubberies and copses, and by wood-walks, where they will grow as freely as any native weeds, and in many cases prove charming in autumn'. A century and a half later both plants have been democratised. No longer the prerogative of those lucky enough to have their own wood-walks, they crowd along the embankments of railways and motorways throughout suburban Britain, a suffusion of autumnal lilacs and golds along these drab routeways.

William Robinson's 'wild gardening' was only one of the influences behind the spread of these exotic plants. But in using overseas weeds as garden plants, and then championing a planting style sufficiently natural to guarantee their spread, it highlighted just how tenuous the boundary was between the province of weeds and the theatres of cultivation. Plants could stray across it not only physically, but conceptually, too.

# 9

# *Grelda*

## The witch in the border

THERE ARE BOUNDARIES everywhere between
nature and culture. One lies about fifty yards from where
I'm writing. The front of our garden in Norfolk ends in
a grass verge that runs along the edge of the lane. It's
legally our property, but has a debatable status, being also
the public margin of the road, visible to passers-by and
useful as an escape route for walkers when there's traffic
about. We've treated it much as if it were any other roadside verge. We enjoy the cow parsley and primroses in the
spring, the fat tussocks of yarrow and plantain later, and
aren't troubled by its muddle of seedheads and bunched
grasses in high summer. I suppose we mow it a maximum
of twice a year, which is standard for rural verges.

But this wasn't enough for one neighbour. One late
summer day we received a letter from the Parish Council
informing us that there had been a complaint about the
untidiness and weediness of our verge, and urging us to
manage it more fastidiously to bring it into line with the
neighbouring frontages. Both civic pride and public util-
ity, it was darkly intimated, were at risk. I sent a rather
haughty letter in response, doubtless overstating my case.
I argued that billiard-table turf might be *de rigueur* on

173

lawns but wasn't appropriate for a country roadside, and that what our anonymous neighbour regarded as weeds were the same wild flowers that grew relatively unchecked alongside the uninhabited part of the road, part of the biodiversity that even the government was exhorting us to conserve. I heard nothing back. Game to me, I thought. When we returned from a holiday a couple of weeks later, the verge had been mowed. Vigilantes had done the job for us. I didn't feel that, in this instance, a suit for damages would get very far.

This was a small border skirmish in the weed wars, but a sharp reminder that feelings about the boundaries between wildness and domesticity not only run high, but are affected by all kinds of subtle social considerations – fashion, community solidarity, class, horticultural fervour. Two years after this episode another botanically inclined vigilante emerged in the village, and over three summer nights burned to the ground almost every local hedge of *Leylandii* cypress. He was probably just a local youth who'd discovered the satisfying flammable nature of conifer foliage. But the single-species focus of the attack was hard to gloss over. *Leylandii* – funereal, light-stealing, unneighbourly – is the most contentious as well as the most popular shrub in Britain. And it is the only cultivated species whose height is specifically controlled by law, as if it were a backyard weed in Houston.

~

In the United States the 'front garden' – though it is not a 'garden' in the English sense – is in the public domain from the outset. Across suburban America the space between house and road is, almost invariably, occupied by lawn. Each property's grass frontage joins seamlessly with the next, and the neighbourhood lawn forms a continuous stretch of grassland often miles long. Across the whole country lawns occupy some 50,000 square miles, approximately the area of the state of Iowa, and householders spend more than $30 billion a year on maintaining them. More chemical weedkiller and fertiliser is sprayed on them per acre than on any other crop in the country.

The pressures to conform to the orthodox standards of lawn perfection are huge. There are no hedges to hide behind. Your tolerance of a tuft of plantain is not just a sign of your own slovenliness, but a public insult to your neighbours. Your lawn is a visible extension of theirs, and of the whole community's proudly maintained estate. It is part of the greater American lawn. If you default on its maintenance, you have opted out of the social contract.

In his book *Second Nature* Michael Pollan tells the story of a dissident lawn-owner in Buffalo, New York. He was a Thoreau scholar, and had turned his frontage into a wild-flower meadow. His outraged neighbours mowed it down, prompting him to erect a sign saying: 'This yard is not an example of sloth. It is a natural yard, growing the way that God intended.' A local judge, citing an ordinance similar to the one in Houston, ruled that his 'wild flowers' were in fact 'noxious weeds', and ordered him to mow them down, or face a fine of $50 a day. He refused, and when the case was last reported, his act of suburban civil disobedience had cost him more than $25,000.

How has a nation that prides itself on being the bastion of individuality come to take such a fiercely collective view on the proper condition – and appropriate vegetable citizens – of its domestic grassland? Pollan suggests that, 'like the interstate highway system, like fast-food chains, like television, the lawn has served to unify the American landscape'. But more diverse products flow through these somewhat limited channels than are ever permitted on the national lawn. A more complicated interweaving of America's historic attitude towards nature and its notions of civic responsibility seems to be involved.

In 1868 the landscape architect Frederick Law Olmsted designed, just outside Chicago, one of the first planned suburban communities in America, and laid out the ground rules for what would become a nationwide institution. Each house would be set back thirty feet from the road, and any kind of exterior division – walls, hedges, fences – was banned. This was partly a dig at the 'high dead walls' of English culture, which he believed made a row of houses seem like 'a series of private madhouses'. The American front garden would be democratic and egalitarian. When Frank J. Scott took up Olmsted's baton in the 1870s, and published *The Art of Beautifying Suburban Home Grounds of Small Extent*, he declared, 'A smooth, closely shaven surface of grass is by far the most essential element on the grounds of a suburban house.' It would contribute to a communal landscape, and 'the beauty obtained by throwing front gardens open together, is of that excellent quality which enriches all who take part in the exchange, and makes no man poorer'. But Scott also sought to make a particular vision of 'nature' part of the communal equation. 'It is unchristian', he asserted, 'to hedge from the sight of others the beauties of nature which it has been our good fortune to create or secure.' The lawn is the ordinary

American's slice of the wilderness. Because it is green it is still 'nature'. Because it is mown and sprayed it is also culture, 'created or secured'. The collective lawn became a vehicle of consensus, a symbolic expression of America's irredeemably contradictory attitudes towards the land: that it is both a great, unfenced, communal resource, liberated from the petty individualism of Europe, and at the same time an arena for the aggressive advance of the pioneering spirit.

The problem is that, once established, a lawn dictates its own standards. It is not just America, or even the watchful neighbourhood, that places immediate pressures on the owner of a lawn. It is also the turf itself, the demands made by its singular, unblemished identity, its mute insistence that if you do not help it to continue along the velvet path you have established for it you are guilty of a kind of betrayal. The sociologist Paul Robbins has coined a term for the suburban victims of the combined pressures of national tradition, neighbourly prissiness, commercial gardening pressures, and the insistent identity, the integrity, of the lawn itself. He calls them 'Turfgrass Subjects'.

But let us indulge in a little travel in time and space, move 800 miles east of Chicago to the small town of Concord in Massachusetts and twenty years back in time from Frank Olmsted's inaugural sowing of the seeds of America's national turf-lot. There the writer who, 150 years later, was to inspire a New York historian to defend the 'noxious' plants desecrating his lawn is thinking hard about a different attitude towards weeds. Henry Thoreau had begun building his one-roomed shack by the side of Walden Pond in 1845. He lived there, growing his own food and living more or less self-sufficiently, for more than two years, garnering the thoughts and experiences that would fill one of the greatest works of American literature. *Walden; or Life in the Woods* (eventually published in 1854) is notionally an account of his two-year experiment. But it is really about what it means to be a citizen in the fullest sense, how to dwell in a place with simplicity and grace, to cohabit respectfully with your fellows, of all species. He wanted to live close to what he called the 'marrow' of life.

And part of the essential marrow of *Walden* is a short and famous essay called 'The Bean-field'. It is sometime during the late spring of 1845, and Thoreau is contemplating his beans. He reckons their rows measure more than seven miles in total length, and he is hoeing them obsessively. He's not sure why, except that he seems to have become, to adapt Paul Robbins, a 'Bean-row Subject'. It is, as he puts it, a matter of self-respect; but also of row-respect. He loves his rows, whose initial order seems to demand continuing order. So, under his merciless hoe, out come the brambles and St John's-wort and cinquefoil, though he is at a loss to find a rational reason for this 'Herculean labor'. He has planted far too many beans and does not even enjoy eating them. His pleasure is in the ritual of the work, even as he comes to understand that he is simply replacing one kind of weed with another:

> Early in the morning I worked barefoot, dabbling like a plastic artist in the dewy and crumbling sand, but later in the day the sun blistered my feet. There the sun lighted me to hoe beans, pacing slowly backward and forward over that yellow gravelly upland, between the long green rows, fifteen rods, the one end terminating in a shrub oak

copse where I could rest in the shade . . . Removing the weeds, putting fresh soil about the bean stems, and encouraging this weed which I had sown, making the yellow soil express its summer thought in bean leaves and blossoms rather than in wormwood and piper and millet-grass, making the earth say beans instead of grass – this was my daily work.

The watchful community, 'sitting at ease in their gigs', pass by the barefoot labourer, gossiping about the lateness of his planting and the dishevelment of his fields. Thoreau is more interested in watching the birds circling above him – nighthawks and harriers 'alternately soaring and descending, approaching and leaving one another, as if they were the embodiment of my own thoughts'. 'When I paused to lean on my hoe, these sounds and sights I heard and saw any where in the row, a part of the inexhaustible entertainment which the country offers.'

The next summer, he decides to abandon beans altogether. They – and their hoeing – had become no more than a habit, a distraction from the more fundamental teachings of the field. The sun, he realises, looks on wildness and cultivation without distinction. 'This broad field which I have looked at so long looks not to me as the principal cultivator, but away from me to influences more genial to it, which water and make it green. These beans have results which are not harvested by me. Do they not grow for woodchucks partly? . . . Shall I not rejoice also at the abundance of the weeds whose seeds are the granary of the birds?'

~

How weeds are regarded in anything but the most intensely commercial plots is always, as it was for Thoreau, a matter of internal debate, as well as a consequence of social and cultural pressures. In one's own garden the status and fortunes of weeds are influenced by personal tastes and prejudices, family tradition, passing moods. They, or their absence, help create the ambience of 'home' every bit as much as the colour of the front door. Gatecrashers may or may not be welcome. The complicated, finicky business of gardening is, at root, almost exclusively about what shall be encouraged in the home patch and what shall be driven out.

Our garden in south Norfolk is no exception. The weed policy my partner Polly and I follow (and don't always agree about) is whimsical and sometimes downright hypocritical. It bows to culinary need and a few social conventions, but is hedged about with sentimentality and a strong sense of the history of the place. I guess the garden was first made around 1600, at the same time as the main timber-framed house was built. The property was a small farm then and sat, conveniently for the owners, at the edge of one of the village commons, where glamorous local wildings such as sulphur clover and spiny restharrow doubtless grew. The earliest detailed map I've been able to find is from the early nineteenth century, and it shows, as might be expected of a smallholding, a strictly functional garden. There are two rows of fruit trees at the front of the house, and a pond at the back. What is now our meadow was the eastern corner of a field labelled 'Hempland' by the surveyors. This was exactly what it says: a plot devoted to the growing of cannabis. The two bachelors who lived there at the time were farming hash ('weed' in a later street slang) on our lawn. Except that this was a non-psychoactive variety cultivated for making cloth. It was the favourite smallholder's crop in this valley. The damp sandy soils echo *Cannabis sativa*'s natural habitat in central Asia, where it was originally an annual weed. Its cultivation was a homely business. After the hemp stems were cut in the summer, they were soaked in the ponds for a week ('retted') to help separate the long fibres from the woody rind. Then they were beaten and 'scutched' with a sharp wooden scraper to remove the rind. Finally, the fibres were 'heckled' – straightened and combed – until they were ready to be woven on Jacquard hand-looms into a superior kind of linen. Kensington Palace and Eton School were on the long list of regular customers for our valley's – and maybe our own garden's – most famous export.

Despite its long and prestigious contribution to local culture, no relic of the crop seems to have survived, at least in our parish. Here and there, as the qualities of hemp fibre are rediscovered, a field of the extraordinary eight-feet-tall plants reappears, usually behind a very tall hedge, which, on warm days at least, completely fails to hide the heady aroma. But in any other situation it's now regarded as the worst sort of weed. It's a toxic alien, and illegal to have about your property without government permission. When I applied to the Home Office for a licence to grow a patch of the less mind-altering variety on the land it had occupied two centuries before – for strictly historical reasons, I stressed – it was made very clear that home-growing was precisely what legislation was meant to prevent. Oddly, the criteria for approval, carefully spelt out, seemed more concerned to minimise the threat of theft than prevent drug abuse. But in them I sensed the tolling of that ancient fear that forbidden plants could contaminate the soul as well as the earth, and should be kept away from vulnerable humans. 'It is for the grower to decide the most suitable location', the guidelines concede, 'but ensuring that it is only grown where there is minimal risk of attracting the attention of those who might steal the crop'. It must not be cultivated 'beside busy public roads or close to housing, industrial or leisure areas [or] where, for example, folds in the land can be utilised to shield crops . . .' How that word 'fold' reverberates with its sense of shepherdly protection.

In the end the lack of a licence proved irrelevant. One warm summer, a single cannabis plant appeared of its own accord in the herbaceous bed, its pale, famously fingered leaves waving mischievously between the phlox and the clarkias. It grew to about three feet high, put out its dull yellow blooms in October and keeled over with the first frost. I'd like to believe that it was a long-dormant progeny of the crop our nineteenth-century bachelors grew for the smart London linen trade. But it was most likely a casual, sprung, like the millet that appeared close by, from birdseed, and a reminder that weeds always find their ways back to places they like.

There is rarely much botanical continuity in a garden. New owners arrive with more modern tastes, beds are redesigned and replanted, plants fall in and out of favour, modish newcomers are introduced from the far corners of the earth. Our garden must once have had a crown imperial growing close to where the cannabis sprouted, to judge from the intense smell of fox that regularly appears in exactly the same spot at exactly the moment in late April when crown imperials bloom. But no plant ever appears. It is a botanical ghost, an olfactory fossil, springing perhaps from some dormant fragment of root. The plants most likely to physically survive the upheavals of history are very old trees and very nimble weeds. The mugwort and small nettle that edge our drive may be direct descendants of the weeds that grew in the hempfield two centuries ago. The groundsel in the lettuce bed may belong to a lineage that goes back 3,000 years to the time that Bronze Age farmers first worked in this valley. As for the sowthistle that sprouts impudently from the thatch, I've no idea of its provenance, but it does cry the weeds' call: *We were here before you, are your constant and ubiquitous companions, and will be here when you are gone.* (There is, as if to prove the point, a specific weed of thatch. The squat and reassuringly fleshy rosettes of houseleek were once planted on thatched roofs as a magical protection against lightning. They've lingered and more or less naturalised there, and, prone on the rooftops, acquired the longest and most cryptic of all vernacular plant names: 'Welcome-home-husband-though-never-so-drunk').

We habitually think of weeds as invaders, but in a precise sense they are also part of the heritage or legacy of a place, an ancestral presence, a time-biding genetic bank over which our buildings and tinkerings are just an ephemeral carapace. I still hoick them up when they get in my way, but it's a capricious assault, tinged with respect and often deflected by a romantic mood. A sense of the antiquity of weeds is also a reflection of how long they have been familiars in one's own life. They turn up at the same time of the year, every year, like garrulous relatives you wished lived just a little further away. They are clock-worts. Their stubborn regularity may be their worst feature from a gardener's point of view, but it's also a reassuring reminder that life goes on.

There weren't many obvious weeds when we first moved in. The garden was tidy to the point of prissiness. It was hoed, mown and pruned almost every week. And the battalions of weedkiller sprays left in the shed showed just how the intruders were kept at bay. In our first summer, with the chemicals consigned to the bin and our energies concentrated on work on the interior of the house, the weeds *de la maison* (the French call weeds *mauvaise herbes*, bad plants) erupted into mischief as mice are supposed to do when cats are away. In the great heatwave that burned through that summer of 2003 it was as if a huge weight had been lifted from their chests. The ground seemed to breathe them out in the heat, like puffs of floral vapour. Scarlet pimpernels studded the gravel, opening their flowers at breakfast time and closing them after lunch. A big, ferny tansy shot up by the side of the oil tank. Green alkanet (from Spain) showed itself to be the happiest colonist in the whole garden and appeared everywhere: in the potato patch, the paths, the pots of herbs. In the rough grass its clear blue flowers stud the white lace of cow parsley like cobalt buttons.

The oddest weed appeared in June, a scatter of mystifying red-ribbed seedlings in the old vegetable garden. By high summer they'd revealed themselves to be a bumper crop of thornapples – a species that has stalked me since those slum-botanising days in the Middlesex badlands. By July their pale blue flowers were blowing foppishly amongst the workmanlike rows of French beans and seedling tomatoes (to which they are related). By late August the wickedly spiny, conker-like fruits had formed. Goodness knows how the plants got here. The commonest source of seed these days is bags of garden fertiliser from South America. But thornapples were anciently cultivated as the source of the alkaloid drugs atropine and hyoscine, used in the treatment of asthma and digestive disorders, and their seeds have provenly long dormancy. It wouldn't surprise me if our hemp-growing forebears had a specimen in their herb patch.

And I suppose it is remotely possible that I'd inadvertently introduced them myself, that a few seeds from a fruit gathered thirty years before were lurking in the corner of a box, or stuck between the pages of a book – maybe against an illustration of themselves. Increasingly I find that weeds are far from happenstance, that we, as users and workers of the garden, are in some way generating them – *cultivating* them, if you like – by our personal affections and behaviour. Many of them are here because of the people we are, with our own histories and hoardings. They reflect the way we dig and mow, the walks we take, the holidays we go on. There is no need to lift a deliberate finger to get them here, just show a kind of forbearance once they have arrived.

How else to explain the arrival of the straggling oddity that spilled over our gravel one summer? It looked familiar, and I lazily assumed that it was a just a rather droopy specimen of the common viper's-bugloss (*Echium vulgare*) that grows in sandy fields close by – until late August, when it began to resemble a nest of blue-flowered snakes. A serpentine habit is *not* how this family acquired its 'viper' tag, which refers to the imagined resemblance between the seeds and a viper's head (*Echium* is from the Greek *echis*, a viper). Ordinary viper's-bugloss is briskly upstanding in its growth. So I decided to look at our gravel scrambler properly, and not just as part of the impressionist haze of summer weeds. Under a lens its pleated blue flowers had short stamens, not protruding beyond the petals – 'included' as botanists say. Common viper's-bugloss has long pink stamens, which seem to poke and flicker out of the flower's mouth – another snaky feature. This was a different species, small-flowered bugloss, *Echium parviflorum*, an annual weed of fields and dry waste places in the Mediterranean. Had we brought a few seeds back on our shoes, after our last holiday tramping the luxuriantly overgrown farmland of Provence? Had they been in a box of figs, or the packing round a pot? Whatever route they'd arrived by, it was almost certainly our lifestyle, not some abstract plant-dispersal mechanism, that brought them to our house. They were bespoke weeds.

Vervain, whose tiny, bright lilac flowers open along its wiry stems like a slow-burning sparkler, may be another. It pops up from time to time in our brick paths, jumps into pots, fizzles between the dwarf beans. It grows on a few sandy roadsides nearby, and may have originated there. But it was one of the Anglo-Saxons' sacred herbs, a magical charm against witchcraft and a supposed cure for the plague in the Middle Ages, and Polly grows it reverently in the Norwich Cathedral physic garden she helps run. The garden tools she uses there are the same ones which turn over our more secular soil, and small clods of earth migrate back and forth between the two gardens. If Sir Edward Salisbury were still around he could probably trace our daily movements from the botanical evidence stuck to our car tyres.

I like this sense of weeds as archaeological artefacts, embodying history as if they were arrowheads or old letters, charting our habits and beliefs. Except that in another sense they are nothing whatever like museum specimens, and are wonderfully and mischievously alive. We were quietly pleased to find a row of winter heliotrope at the foot of our hedge, next to where the cars are parked. It's a loathed weed of roadsides in some parts of the country but has an intriguing pedigree and a glamorous disposition. The plant was apparently unknown in Europe until the late eighteenth century, when it was discovered in France, growing at the foot of Mont Pilat in the Massif Central, its spikes of pale lilac, winter-flowering, scented tassels crowding over horseshoe-shaped leaves. It was taken up by the Parisian aristocracy to grow in pots in their winter gardens, and reached Britain in 1806. The name heliotrope ('sun-following') is misleading. The flowers of the winter species don't rotate to track the sun as do, say, sunflowers; but they have the same seductive marzipan-and-vanilla smell as the true, summer heliotrope (*Heliotropium arborescens*, from Peru), which gives the plant its other name of cherry-pie. They first appear just before the winter solstice, bringing to the year's least-scented days a fragrance that seems to belong both to Christmas and the coming spring. When I first moved to Norfolk, and was lodging alone in a big sixteenth-century farmhouse, I kept a vase of them on the bookshelf the whole winter.

But taken out of its Mediterranean constraints and planted up in the richer soils of the north, heliotrope ramped away. It was thrown out of most gardens for bullying, and took up residence on damp roadsides. It spreads in big clonal patches (some of which do not even flower) and builds up tesserae of evergreen leaves that act like a smothering tarpaulin over other low flowers. In a bleak December though, with not another wild flower about, it can touch your heart.

And I would feel deprived if I had no greater celandine in the garden. This rather ungainly yellow-flowered weed, brought here anciently from the Mediterranean, was the species that introduced me to the labyrinthine world of plants' cultural connections. It was the coincidence between its name and that of the quite unrelated lesser celandine that first intrigued me. 'Celandine' comes from *khelidon*, the Greek for 'swallow', and the greater celandine (*Chelidonium majus*) may have been so-called because it flowers at the time of the swallows' return. John Gerard wasn't convinced, and suggested that it was 'because some hold opinion, that with this herbe the dammes [female swallows] restore sight to their young ones when their eies be out: which things are vaine and false'. I tracked this belief back to the medieval herbalists, who certainly didn't think it vain and false, and recommended the herb to clean away 'slimie things that cleave about the ball of the eie'. (It was a drastic remedy. The orange latex exuded by the plant is so corrosive it was also used to cauterise warts.) I heard about a possible representation of the plant carved on the shrine to St Frideswide in Christ Church Cathedral, Oxford, which dates from 1289, and found its lobed leaves, unmistakable amongst the equally realistic carvings of sycamore and hawthorn and ivy. Their presence may not be purely decorative, or coincidental. St Frideswide, as well as being the patron saint of Oxford, was a benefactress of the blind. She was the daughter of a twelfth-century Mercian princess, and went into hiding to avoid an arranged marriage. Her luckless suitor subsequently went blind and, in an act of contrition, Frideswide retired to a nunnery. Not long after, she summoned a holy well to spring up in the village of Binsey, just upriver from Oxford. Its water was reputed to have special powers for eye disorders, and this seems to have been the reason for her elevation, and the presence of greater celandine, a prime eye-herb, on her shrine.

Greater celandine has, aptly, become one of Oxford's signature weeds. I've seen it at the edge of car parks, on the city's venerable walls and sprouting at the foot of exclusive college staircases. And I expected to find it when I eventually made the trip to Binsey. The well was still there, half hidden at the bottom of some mossy steps behind the village church. But there was no greater celandine – just a little clump of the lesser variety, which is in flower two months before the coming of the swallows, and which may share *Chelidonium*'s Anglicised name for no other reason than the colour of its flowers.

Such whimsical fancies can set off reverberations far beyond your own garden, a domino effect in weeds. I like the oddity of the double-flowered greater celandine, and when I was living in the Chilterns, once pilfered a seed-pod from the fulsome specimen that grows in the Queen's Garden at Kew. It germinated without any problems in my own herb garden, and the following year sprang up also in some cracks in our concrete sideway. Two years later it was next door, and by the time I moved to Norfolk I could follow a track of double celandines a quarter of a mile long, all the way down our street, across the main road and into a factory car park, where they ran up against a high wall and stopped. There are single-flowered specimens popping up in our Norfolk garden but they seem to have none of the adventurous vigour of the high-bred double.

Most 'new' weeds begin as garden escapees. In what can feel like the ultimate act of horticultural ingratitude, your own garden plants can even become rampaging weeds within the bounds of your own garden. Their expansiveness makes a travesty of your subtle planting. They scale walls, seed *into* walls, give the neat rectangles of the vegetable beds the raddled brilliance of a cubist painting. The small corms of montbretia get everywhere, rooting sometimes at the edge of bonfires where we'd torched the root-balls, hoping to exterminate them. The mints have invaded the lawn. The lawn itself has become a virulent weed, encroaching remorselessly on any adjacent patch of ground we'd rather devote to something else. And so successful has been my attempt to make a facsimile Mediterranean garden with a couple of tons of limestone rubble that the species I've lovingly nursed through frosty winters have become invasive to each other, and I spend more energy weeding out oregano and euphorbia seedlings than I do the docks.

One species that I introduced to our wild-flower meadow to control what I regard as its 'weeds' – rank grasses like couch and rye-grass – has turned into an aggressive colonist itself. Yellow-rattle, so named because its seeds rattle about in the inflated pods when they're shaken by the wind, is a partial parasite. It has green leaves, and makes some of its own food. But its roots have suckers which attach themselves to the roots of grasses to draw off nutrients, sapping their vigour in the process. Wildlife gardening manuals urge you to sow it in potential meadows, to keep the grasses in their place and give showier wild flowers more room. What they don't tell you is that yellow-rattle is also parasitic on many other species. Its promiscuous habits were the subject of my friend Chris Gibson's PhD. He disentangled the plant's rootlets' progress through the turf, inch by inch, and found that it preyed on at least a dozen different plant families. In our meadow it has reduced clovers and vetches to dwarves along with the grasses. In places it is so dense that there is no grass visible at all, and I wonder if it had begun parasitizing itself. But no ecosystems are stable. The terrible cold of early 2010 encouraged massive germination of rattle seeds. Then, during the long drought that followed, patches of already weakened grass began dying, promptly followed by their hangers-on. By midsummer the meadow was dotted with patches of desert, already being colonised by more traditional pioneer weeds. Rattle may seem nothing more than a freeloading nuisance, but it can also be a creator of biodiversity.

But we find ways of tacking with our weeds. It simply needs a change of perspective, what psychologists call 'reconfiguring'. So the opium poppies and musk-mallows that occasionally spring up amongst the potatoes are left to their own devices, and redefined as ornamentals. They do no harm, and brighten up what is otherwise the dullest patch of the vegetable beds. When hedge bindweed climbs the hawthorns and drapes then with showy white trumpets, it's welcomed – and I defy anyone to put a hand on their heart and say it is less beautiful than morning glory. But when it ramps through the cistuses we yank it out it in long, surrendering lianas – a deeply satisfying task, because unlike balls of string, bindweed runners are never naturally knotted (though they will take a knot, and Polly, ever the improviser, uses them as makeshift garden twine).

And weeds are the very stuff of life for insects. Brimstone butterflies gather nectar from early buttercups. The caterpillars of small tortoiseshell, peacock and red admiral feed on nettle leaves. And to the question, 'What are weeds for?' one answer might be, 'Moths.' I have not seen three-quarters of these species, but the list of moths that feed on that governmentally scheduled weed, the dock, reads like a found poem: bearded chestnut, black rustic, blood-vein, brown-spot pinion, chestnut, common marbled carpet, cream wave, dark-barred twin-spot carpet, dark chestnut, feathered ranunculus, garden tiger, gem, green arches, grey chi, Isle of Wight wave, large ranunculus, large twin-spot carpet, Lewes wave, mottled beauty, muslin moth, nutmeg, pale pinion, Portland ribbon wave, red sword-grass, riband wave, ruby tiger, satin wave, striped hawk-moth, sword-grass, twin-spot carpet, white-marked, wood tiger, yellow shell.

It's on the lawn that I suppose we make most impact on weeds, not because we prise them out, but because we mow off their tops along with the grass. This, of course, by the iron rule of weeds, means that we give the edge to all those species that relish beheading, and have evolved leaves that grow flush with the ground. So the lawn is dense year-round with rosettes of plantain and dandelion. From as early as January, long before mowing begins, the first daisies are studding the grass (spring has arrived, they say, when your foot can cover three, or seven, or a dozen – the number varies even between next-door neighbours). In March the first ground-ivy flowers bring trails of blue and purple below the level of the grass. Where a beech tree shades a part of the lawn they replace the grass altogether for a couple of months with a pool of metallic blue.

But the best lawn weed, the flower that says, decisively, here is the spring and the new sun, is the lesser celandine. It's rather fussy in our garden, and only really flourishes in a damp corner under the cherry-plums which we mow no more than three or four times a year. But for six or so weeks from the middle of February it makes that dappled glade *shine*. It's the only word. Celandine's petals, like buttercups, seem able to reflect the light, as if they were made of yellow metal, or oil, or most persuasively, molten butter. John Clare tells of a game where children held the open flowers under each other's chins (as buttercups are today) to see if a golden reflection foretold a golden future. One modern Warwickshire child, mishearing the plant's name, called it 'lemon-eye'.

The flowers register the sun uncompromisingly, opening wide on warm days and closing up in the cold. In Dorset celandine was known as 'Spring messenger', straight and simple. When I was young I used to try and pick a bunch of the opened flowers each year as a Valentine's Day posy, and one cold February had to force the buds open with a sun-ray lamp. Wordsworth noticed its precocious flowering, and wondered why such an exquisite bloom had not been more feted. In a short note before one of his own poetic tributes to the weed, he says: 'It is remarkable that this flower, coming out so early in the Spring as it does, and so bright and beautiful, and in such profusion, should not have been noticed earlier in English verse. What adds much to the interest that attaches to it is its habit of shutting itself up and opening out according to the degree of light and temperature in the air.'

For those of us who share Wordsworth's view, it is mysterious why celandine is hounded from most lawns, and why a turf of pure velvet green is preferred to a multi-coloured quilt. But of course it is not a simple matter of aesthetic preference. Most gardeners, I suspect, enjoy celandines and daisies and speedwells. But preferably somewhere else – in a meadow, on a roadside, tucked into a child's posy. The lawn in England may not be the badge of social conformity that it is in the United States, but it does generate its own intrinsic standards, different from other kinds of grassland, and the intrusion of a wild flower moves it towards another category, the meadow. The wild flower, like all plants that slip into the wrong category, or wrong place, then becomes a weed by definition.

But there are species whose pedigree stops all but the most single-minded mowers in their tracks. Our native orchids share a common ancestry with some of the most beautiful blooms on the planet, a family of glamorous habits and exotic origins whose high-bred descendants are one of the staples of modern florists. They have good looks, rarity value and an unbeatable provenance. Once identified they would be the last candidates for anybody's weed list. But two of our native species prefer very short grass, and therefore often invade lawns. So the stage is set for a classic collision: the irresistible intruder meets the unshakeable category.

Autumn lady's-tresses is a denizen of the southern chalk country. It's chiefly found on downland that has been grazed short by sheep or rabbits, but often springs up on old lawns. I've seen a tennis court in Kent covered with its slender stalks and diminutive white flowers, which might not even warrant a second glance, but for the fact that they're arranged in a spiral round the stem, and smell of jonquil.

But nobody would overlook a bee orchid. I remember seeing my first on an evening picnic on a Chiltern chalk-down, and feeling that I had gone through some subtle graduation in the rites of botany. It was not just the fantastical look of the flowers, the chimerical sense of a pink fairy's wings joined to a brown bumble bee's body. They seemed to transcend the realm of the vegetable altogether, to be ornaments of porcelain and velvet that had been mysteriously animated by the sun. They challenged my whole human-centred and inexperienced sense of what a flower should be like.

I am not the only organism to have problems with the bee orchid's ambivalent identity. According to evolutionary biologists the bloom developed its extraordinary form to deceive real bees. In theory the bee mistakes it for one of its own, engages hopefully in what is called 'pseudo-copulation', and inadvertently picks up pollen on its legs, which is transferred to its next port of pseudo-copulatory call. The trouble is that the flower refuses to play its Darwinian role. It's entirely self-pollinating, and in Britain no one has ever glimpsed a bee attempting so much as the first moves of foreplay.

Bee orchids are equipped with some familiar weed technology. The flowers produce many thousands of dust-like seeds, which can be blown great distances on the wind. If they come down on disturbed chalk soils they can build up huge colonies of flowering stems. Old quarries are the classic bee orchid location. More conspicuously they've appeared in numbers on the spoil tips of chemical factories, on a new roundabout outside Hitchin, at the edge of the car park at the Milton Keynes Telephone Exchange and on the running track of a smart private school in Oxfordshire.

Bee orchids take up to eight years to grow from seed to flowering. But in their third or fourth year the seedling develops a rosette of leaves as flat as a plantain, which is not only immune to light mowing, but actually benefits from the removal of competing leafery. We had a colony spring up under our washing line, where I tended to keep the grass no more than an inch and a half tall. I noticed the rosettes in January, not entirely sure what species they were. They were unmistakable by May, and in full extravagant bloom by mid-July. I guess the seed may have blown in from our nearest colony, half a mile away in a typical bee orchid hide-out: the sandy surrounds of an electricity sub-station, used by local kids as a bike scrambling track.

They were our proudest sprouting that year, the prize blooms we showed off to visitors long before the old roses. But to those gardeners in thrall to the unyielding definition of lawn, bee orchids pose a tricky dilemma. I was once contacted by a lady back in the Chilterns because a large number of mysterious growths had appeared on her lawn. From her description they sounded like bee orchids to a T. When I went to investigate I counted more than a hundred in full flower on her small patch of grass. Except that somehow (perhaps with an electric razor: it looked that surgically precise) she had tight-mown the grass *between* them. The orchids, shorn of their leaves in the putting-green turf, looked like tiny plastic windmills.

But there is one weed species that is beyond the pale even

under our laissez-faire regime. Ground-elder doesn't trouble the vegetables, and is absent from the lawn. But in the
herbaceous borders it permeates every inch of soil. The
patches don't just occupy the places between the cultivated flowers. They subvert them, insinuating their white
subterranean tendrils, as supple as earthworms, around
and through any root system in their way. Ground-elder
is almost immune to weedkillers, and hard to deal with by
ordinary hand-weeding. Any fragment of root left in the
ground generates a new tuft. It is a mystery why it hasn't
moved out from gardens (it would only need to travel fifty
yards from ours) to become one of the most ineradicable
weeds of agriculture.

Given its behaviour and taste for disturbed ground,
ground-elder is usually assumed to be an alien, an early
introduction from continental Europe. No evidence of the
weed's presence in Britain has been found in prehistoric
excavations, though it begins to appear in Roman deposits. The Romans certainly valued it both as a medicinal
herb (chiefly for treating gout) and as a vegetable, and it's
probable that they introduced it along with fennel and
alexanders, two other kitchen-garden species which rapidly naturalised. It soon became familiar enough to pick up
local and vernacular names. Goutweed is self-explanatory,
as is its variant bishopweed (their eminences being pro-
verbially prone to that ailment). Jack-jump-about appears
in the sixteenth century, an ominous pointer to a pattern
of behaviour which was later confirmed by John Gerard,
in untypically desperate tones: 'where it hath once taken
roote, it will hardly be gotten out againe, spoiling and
getting every yeere more ground, to the annoying of better herbes'. 'Getting every yeere more ground' is a precise

description of ground elder's expansiveness. In a single summer each underground stem can extend as much as three feet, so that from a single rosette an area of more than a yard square has been colonised. The roots plunge down as well, to unprecedented depths. In a quarry in Kent in the 1990s, a worker found ground elder roots still probing thirty feet below the surface.

Its method of spreading is similar to those of nettle and bindweed. Networks of underground stems bearing new shoots ramify under the ground, penetrating not just the vacant soil, but any gaps in other root systems. Each stem can extend up to three feet in a season, terminating in a few leafy offshoots, some of which, to the vexation of gardeners who wish all weeds were ugly enough to hate without reservation, produce umbels of beautiful creamy white flowers in June. And, like bindweed, almost any fragment of the wiry root system or shoot, cut up by hoe or spade, can generate a new plant.

Our wide herbaceous border was thick with ground-elder when we first moved in. Constant hoeing and hand-weeding reduced its visible presence, but more would appear just weeks later. It has been the bane of Polly's gardening life, and she has made an inspired contraction of the name of this seemingly immortal witch-weed, to focus her hostility. She calls it Grelda.

In the end more drastic action was required to limit its presence to tolerable levels. We decided to halve the width of the vast perennial avenue, to reduce the work of ground-elder weeding as much as anything. All the existing flowers were dug up in clumps and put aside, and Polly began a process of purification that made Thoreau's 'Herculean labor' in his bean-patch seem like window-box preening. She dug and raked the soil to a depth of two feet and sifted it to take out the little white tendrils. She soaked every flower root in water, and worked it over with a small knife and fork until all the imbricated roots were removed. The entire vermicular harvest was then consigned to the bonfire, and the flowers replanted.

The following spring there was scarcely a ground-elder to be seen. But rejoicing at a comprehensive victory may be premature. Here and there, near the edges of the bed, I found a few small leaflets unfolding. I carefully extricated the plants, trying not to break the roots, to see what they'd emerged from. Each seemed to have sprung from a small section of cut root, thin enough to have been overlooked in the great cleansing. And the new shoots were growing, not from nodes along the serpentine root fragment, as in some other species, but from a bulbous swelling at its tip. Under the microscope the whole growth-point, the launch-pad for the next generation, looked like nothing so much as a questing spermatozoon, Grelda's other half.

# 10

# *French Willow*

## Fireweeds

'IT WAS LIKE AN ENCHANTED LAND,' wrote the war artist Sir William Orpen of his first sight of the poppy-blazed fields of the Somme. From our standpoint it is hard to credit, but in those rare moments when the battlefield wasn't an unequivocal vision of hell, British soldiers glimpsed in it a kind of wild garden. It was as if the weeds offered a glimpse, against all the evidence, of the indomitability of life. Captain Ted Wilson, a 29-year-old schoolmaster, wrote to his mother about the colours of the trenches in late spring:

> between the town and us, is a village, which is quite a ruin – its church spire a broken stump, its house walls honeycombed with shell fire. Then comes a great blazing belt of yellow flowers – a sort of mustard or charlock – smelling to heaven like incense, in the sun – and above all are larks. Then a bare field strewn with barbed wire – rusted to a sort of Titian red – out of which a hare came just now, and sat up with fear in his eyes and the sun shining red through his ears. Then the trench. An indescribable mingling of the artificial with the

natural. Piled earth with groundsel and great flaming dandelions, and chickweed, and pimpernels, running riot over it.

If the letters home of the officer class often show a kind of existential thrill in the vividness of the battlefield, the ordinary soldiers took a more tender view. Most of them were country boys, and the wild flowers that sprang out of the morass were the same ones they knew from their home fields. They became the currency of one of the oddest and most poignant attempts to catch an echo of England amidst the horror: trench gardening. An ex-editor of *The Garden* wrote home about how weeds like celandine and cuckoo-pint were transplanted from the surrounding fields and ditches into little plots alongside the trenches, and edged with scraps of battle debris. One private described how a trench just forty-seven yards long had been decked up with 'basketwork and trellis . . . Bottlenecks and junctions had a homely atmosphere with nasturtiums climbing the trellises.'

Yet if the battlefield was a metaphorical wild garden to some eyes, it was also actual farmland, and the battle itself a grotesque parody of the rites of cultivation. Captain Ivar Campbell wrote home in 1915: 'Looking out over the country, flat and uninteresting in peace, I beheld what at first would seem to be a land ploughed by the ploughs of giants.' The poet John Masefield sent his wife a bitter, matter-of-fact description of transmuting a real French farm to compost: 'we blew the farm & the bricks & the pond & most of the dungyard & all the trees and all the fields to dust & rags & holes, till this is all that can be seen . . . corpses, rats, old tins, old weapons, rifles, bombs, legs, boots, skulls, cartridges, bits of wood & tin & iron & stone, parts of rotting bodies & festering heads lie scattered all about'. The Great War poets, mindful of how men had been persuaded to fight 'in order to preserve and somehow possess the beauties of the English countryside', saw the irony involved in destroying the similar rural landscapes of northern France. Ivor Gurney witnessed France – 'a darling land [blessed with] a mellow and merciful spirit founded on centuries of beautiful living' – being turned into a wasteland. Edmund Blunden took the agricultural analogies and resonances to their logical conclusion, and in his poem 'Rural Economy' presented the war as a farmer, sowing seeds of iron and manuring them with flesh and blood.

> Why, even the wood as well as the field
> The thoughtful farmer knew
> Could be reduced to plough and tilled,
> And if he planned, he'd do;
> The field and wood, all bone-fed loam,
> Shot up a roaring harvest home.

The harvest home was, to start with, a cornucopia of weeds. William Orpen visited the Somme battlefields just six months after 415,000 men had been killed there, and was transfixed by what he saw. In his memoir *An Onlooker in France 1917–1919*, he writes:

> Never shall I forget my first sight of the Somme in summer-time. I had left it mud, nothing but water, shell-holes and mud – the most gloomy, dreary abomination of desolation the mind could imagine; and now, in the summer of 1917, no words

could express the beauty of it. The dreary, dismal mud was baked white and pure – dazzling white. Red poppies, and a blue flower, great masses of them, stretched for miles and miles. The sky was a pure dark blue, and the whole air, up to a height of about forty feet, thick with white butterflies. It was like an enchanted land; but in the place of faeries, there were thousands of little white crosses, marked Unknown British Soldier.

But it was not just weeds that prospered. There was a more profitable harvest on the Ypres Salient in the summer of 1915. The French farmers had reclaimed their land within weeks of the battle moving on. They filled in the trenches and shell holes, and began ploughing. The crops came up more prolifically than had been seen in living memory. *Country Life* was moved to comment on the paradoxical luxuriance of weeds and wheat in what had been a war zone just months before: 'When the tide of battle has surged forward, it has been noticed by many correspondents that the earth, as though in haste to conceal the desolating effects, has produced with strange and prodigal abundance.' The magazine tactfully denied that 'the red rain of battle' – i.e. dead British soldiers – could have made any contribution to the fertility, and instead put responsibility firmly in the enemy camp: 'Chemical fertilisers and munitions come largely from the same source . . . the explosives contain large quantities of nitric acid or nitrates and potash . . . By a grim irony, then, the fertilisers which were abstracted from German agriculture to be used in killing the French have had the effect of fertilising the fair fields of France.'

~

It was the poppies whose brilliant resurgence most deeply touched those who saw them, and gave rise to a symbol which still resonates nearly a century later. Poppies had been ancient symbols of life and death. When the prophet Isaiah refers to them (40: 6–8) it is their transience he plays on: 'All men are like grass, and all their glory is like the flowers of the field; the grass withers and the flowers fall.' But outside the Christian tradition they more often symbolised fertility and new life, the blood-red petals expressing the life force of the earth. And in an exact prefiguring of the mythology of the Somme, the poppies which emerged after the battlefield of Waterloo was ploughed were said to have sprung from the blood of the slaughtered soldiers.

But during the last years of the nineteenth century a more sentimental aura began to settle around the plant. For a while it was the defining ingredient of what came to be called 'Poppy-land' – a stretch of the north Norfolk coast near Cromer that became immortalised by the *Daily Telegraph*'s drama critic Clement Scott. Scott had taken to visiting the area in the 1880s, staying at a local miller's house away from the tourist frenzy on the coast. He fell for the miller's daughter, and for the local landscape, which seemed to him to reach an epiphany in the scarlet flowers that lined the fields and footpaths, and swept down to the very edge of the cliffs. He began to write rhapsodic columns about his 'Poppy-land', presenting it as an Arcadia still governed by the ancient rhythms of the farming round. Poppy-land quickly became fashionable, and the urban tourists that Scott had been trying to escape poured up to the little villages on what the Great Eastern Railway

Company rapidly renamed 'The Poppy Line'. (It still runs under this name.)

Scott's most famous tribute to Poppy-land was a poem set in the graveyard of the clifftop church at Sidestrand, which eventually became a best-selling popular song entitled 'The Garden of Sleep'. It is excruciatingly sentimental, and relies partly on the common confusion that corn poppies possess the same soporific powers as opium poppies. But with our privileged knowledge of what was to happen thirty years later it is hard not to read it as some terrible premonition.

> On the grass of the cliff, at the edge of the steep,
> God planted a garden – a garden of sleep!
> 'Neath the blue of the sky, in the green of the
>     corn,
> It is there that the regal red poppies are born!
> Brief days of desire, and long dreams of delight,
> They are mine when my Poppy-land cometh in
>     sight.
> In music of distance, with eyes that are wet,
> It is there I remember, and there I forget!
> O! heart of my heart! where the poppies are
>     born,
> I am waiting for thee, in the hush of the corn.
> Sleep! Sleep! . . .
>
> In my garden of sleep, where red poppies are
>     spread,
> I wait for the living alone with the dead!

It's possible that Scott's song, with its tug at the heartstrings' memories of lost love and English summers, was still remembered in that other garden of sleep three decades on. But probably not by John McCrae, the man whose own verses sparked off the idea of Poppy Day. McCrae was born in Canada in 1873, was educated at Toronto University and went on to work as a physician. At the outbreak of the war he volunteered for the Canadian Army Medical Service and, with the rank of Lieutenant-Colonel, was on the Western Front by the end of 1914. In May the following year he sent his mother a vivid description of the sound and colour of an artillery exchange: 'Then the large shrapnel – air burst – have a double explosion, as if a giant shook a wet sail for two flaps; first a dark green burst of smoke; then a lighter yellow burst goes out from the centre, forwards.' He also, evidently, registered the red burst of the poppies, and what they represented.

In December he was treating casualties from the second disastrous Ypres offensive, and recalled what he had seen and felt in a poem, which he sent anonymously to *Punch*. The magazine printed it on 8 December 1915.

> In Flanders fields the poppies grow
> Between the crosses, row on row,
> That mark our place; and in the sky
> The larks, still bravely singing, fly
> Scarce heard amid the guns below.

> We are the dead. Short days ago
> We lived, felt dawn, saw sunset glow,
> Loved and were loved, and now we lie
> In Flanders fields.

> Take up our quarrel with the foe:
> To you from failing hand we throw

The torch; be yours to hold it high,
If ye break with us who die
We shall not sleep, though poppies grow
In Flanders fields.

The poem was reprinted round the world. In the United States, a YMCA worker called Moina Michael was so moved she vowed to wear a poppy for the rest of her life. In November 1918, ten months after McCrae had died (of pneumonia) in the Boulogne Allied hospital he was commanding, an acquaintance of Michael's set up a scheme to manufacture cloth poppies in France, and sell them in aid of refugees returning from the war-devastated areas. When the British Legion was formed in 1921, she persuaded it to adopt her poppy project, and when the first Poppy Day was held that November the Legion used poppies made in France. The sale raised £106,000 and the Legion immediately made plans to switch the manufacture to Britain. Nearly a century on, the last combatants of the Great War have died, armed conflicts continue to smoulder across the globe, but poppies are still worn to remember that 'war to end wars'. Remembrance Day poppies may represent a triumph of hope over experience. But in a world full of ephemeral and synthetic brand images, they remain one of the most durable of all symbols drawn from nature.

∼

The devastation of the Second World War nurtured poppies too, but its iconic weed was rosebay willowherb, which unfurled liked a purple surf across the bombed-out areas of Britain's big cities in the summers after the Blitz. It was christened 'bombweed' by Londoners, most of whom had never seen the plant before.

This was the second time in 300 years that the capital had been almost burned to the ground. And the second time that the ruins had been colonised – defiled or blessed, depending on your point of view – by a fireweed, a botanical phoenix. The Great Fire of 1666 had created a paradise for weeds in the medieval heart of the old city. The earth was scorched free of existing vegetation. Films of soot from charred timbers had washed into the crevices of collapsed buildings. Cellars and sewers, rank with moist nutrients, were open to the sun. The plant that arrived to grace the ruins, apparently in prodigious quantities, was a modest member of the mustard family with golden-yellow flowers in the shape of a tiny cross. It was christened 'London rocket'.

The name was notionally botanical: a corruption of *eruca*, the Latin name for a group of peppery-tasting plants in the cabbage family. It must have seemed an apt tag for this mysterious interloper, shooting up in the wake of the flames. But despite its vernacular name, it wasn't a born-and-bred Londoner, but an immigrant from the rocky hills of the Mediterranean. Few in seventeenth-century London knew this, of course, and its sudden appearance in the city was a source of wonder and speculation. One commentator, the botanist Robert Morison, concluded that 'these hot bitter plants with four petals and pods were produced spontaneously without seed by the ashes of the fire mixed with salt and lime'.

But seed there must have been. Perhaps the plants had been around in London, in inconspicuous numbers, all the time, and simply weren't noticed until conditions similar to their Mediterranean homeland produced an explosion of blooms. Perhaps the seeds had come over with foreign hay, and been blown about the city with the ashes from burning stables. The biggest concentration, for some reason, was around St Paul's Cathedral.

Rosebay willowherb was scarcely better known than London rocket at the time of the Great Fire, and certainly not as a rampaging urban weed. John Gerard knew that it grew as a rare woodland plant in the north of England, and got hold of some Yorkshire seed to grow in his London garden. (He thought it highly ornamental, and gave what is still one of the best descriptions; see p. 91.) The early, largely eighteenth-century records all portray rosebay as a rare plant of upland rocks and shady woods. The first record for Northumberland is especially telling: 'Among the rocks and bushes under the *Roman* wall on the west side of *Shewing-sheels*, and by *Crag Lake*. On the banks of *South Tyne* by *Slaggiford*, in Knaresdale, plentifully . . . It is introduced into some of our gardens under the name of French willow; but being a great runner, it makes a better figure in its more confined situation among the rocks than under culture . . . It is reputed a rare plant.' This is not the only reference to rosebay behaving quite differently in cultivation from in the wild. (And notice that tag 'French willow', with its double-edged tribute to a plant whose elegance – and invasiveness – were not quite English.)

Rosebay's reputation for shyness survived into the nineteenth century. In W. H. Coleman's *Flora of Hertfordshire* (1848) it is put down as a rare denizen of woods on sandy soils. At about the same time, one of the women of the Clifford family in Frampton on Severn in Gloucestershire,

who made an exquisite and normally reliable collection
of paintings of their local flowers, produced a meticulous
illustration of rosebay, perfectly catching the carmine colour and pleating of the petals – but misidentifying it as
'Great hairy willowherb or Codling and Cream'. And in
Hampshire, the Revd C. A. Johns, in his classic *Flowers of
the Field* (1853), described 'the Rose-bay, or Flowering-wil-
low' as 'not often met with in a wild state, but common in
gardens'. The Elizabethan and Victorian rosebay seems to
be physically identical to the weed that draped itself across
the ruins of London. But its behaviour was quite different.
It's described repeatedly as retiring, rare, a wraith of the
shade. What happened to turn this shy woodland flower
into one of the most successful urban weeds, which by the
late twentieth century would be adorning every urban car
park and railway embankment in Britain?

The change began in the last quarter of the nineteenth
century. Botanists noticed that the plant was spreading,
seemingly, as Oxford ragwort had, along the vectors of
the railway system. During the First World War its population exploded. It began appearing in huge colonies in
areas of woodland that had been felled for the war effort.
It suddenly became clear why, across North America,
this same species is called 'fireweed'. Wherever forest is
cleared, and the brush burned, sheets of rosebay blossom
the following summer. By 1948, in Gloucestershire, where
rosebay had been a confusing rarity just fifty years earlier, the editor of the county flora, H. J. Riddelsdell, had
little doubt about what was happening, or that aesthetic
appreciation ought to be tempered now that the plant's
population had exploded. 'This species has spread with
great vigour since about 1914, owing to the clearing of

woods . . . The seed is easily carried, of course, and the railway has been a great agent in its spread. Beautiful as the plant is in its flowering season, when it is in seed it creates desolation and ugliness over the whole area.' In the same county in the 1960s, Edward Salisbury remarked that 'I have seen a Gloucestershire woodland in early September as though in a summer snowstorm with the multitude of plumed seeds that appeared in the air'.

In the 1990s a correspondent from Buckinghamshire informed me that 'the local huntsman does not like letting the hounds go through stands of this when cub-hunting as the downy seeds get up his hounds' noses so that they are unable to smell'. The plant once described by John Gerard as 'goodley and stately' was being redefined as a public nuisance.

But was it the same plant? Native species don't usually change their behaviour and tastes in habitat so dramatically and so quickly. A rapid colonisation of disturbed ground is exactly what you would expect of a newly arrived and invasive alien. In the standard *Flora of the British Isles* (third edition, 1989) Professor Tom Tutin speculates that there may in fact be two strains of rosebay – variety *macrocarpum*, the retiring denizen of woods and rocky uplands, and var. *brachycarpum*, a strain from Canada or Scandinavia which thrives on disturbed and burned ground, and which had perhaps found its way to Britain with imported timber in the nineteenth century (though this doesn't explain the invasive behaviour shown in the eighteenth century by the plant once it was transplanted into a garden). But he could not find any consistent physical differences between plants from the different habitats. More recent DNA analysis has failed to find any difference at the genetic level either.

However, recent research on the mechanisms of evolution is revealing adaptations which are not traceable to individual genes. It's long been known that many plants – e.g. juniper and fat-hen – can exist in different forms in different habitats without there being any discernible genetic variation between the types. It now looks as if these 'epigenetic' effects can be produced in individual plants within a very few seasons or generations, by a process as simple as transplantation. Some of this adaptive behaviour is controlled by master gene complexes which are both very ancient and occur right across the living world. The large, aggressive, 'weedy' rosebay may in fact be the original form which developed in open and disturbed post-glacial conditions, and the smaller, daintier form an epigenetic adaptation to shade and woodland. The ancestral form was 'switched on' again when humans created facsimiles of the flower's original home.

I wrote earlier that there are no superficial kinship connections between weeds, which can emerge from almost any plant family. But the latest findings in evolutionary genetics suggest that there may be fundamental gene complexes shared by many weed species, which predispose them, for example, to fast growth and adaptability. Ruskin's quirky and wilfully anti-scientific remark that a weed was 'a vegetable which has an innate disposition to *get* into the wrong place' might yet turn out to be vindicated by molecular biology.

But rosebay's personality change aroused suspicions about its provenance. When it unrolled its provocative purple bunting across the ruins of Britain's cities, no one went so far as to suggest it had been dropped by the Germans. But the public reaction was not the unequivocal delight ('new life from the ruins!') that we like to project back from our romantic and ecologically aware perspective half a century later. Rosebay was growing where people had once lived and that was hard to swallow. A contributor from Sheffield to the BBC's oral history archive *WW2 People's War* recalled his own feelings (and, typically, how the plant was seen not as native but a trespasser from the garden):

> These wild specimens of Rosebay are to be regarded an escape from cultivation. It looks well in the garden but the vigorous creeping roots make it a nuisance in a confined space . . . Looking back it always seemed that the summers were hot and the small parachute-like seeds of the Rosebay would blow on the hot summer breeze. When walking through the town centre my mother would keep a tight hold on my hand and with my other hand I would have to ward off these seeds. Inevitably I would get one in my eye and I would try to remove it with my free hand. My mother, preoccupied with her shopping, would say 'Stop rubbing your eyes, you will make them sore.' I hated those derelict scars of land that cut through the centre of my home city and the Rosebay Willow Herb that lurked there.

In London the most perceptive account of the wild greening is a fictional one, Rose Macaulay's novel *The World My Wilderness* (1950). It is set in London in 1946, the year after the war ended. But its spark seems to have been lit five years earlier. On 10 May 1941 Macaulay's flat, including all her books, was totally destroyed in a bombing raid (which also levelled a sizeable section of the British Museum). From then on a sense of disorientation and inconsolability runs through her work. She writes to friends: 'I am bookless, homeless, sans everything but my eyes to weep with.' The experience soon found some catharsis in a short story called 'Mrs Anstruther's Letters' about a woman who also loses all her books – and all the letters from her lover.

Macaulay's real healing began with the gift from publisher Victor Gollancz of a complete *Oxford English Dictionary*. 'My OD was my Bible, my staff,' she wrote, the word not exactly becoming flesh, but at least a reviving spirit. But at the same time she was engaged in a more earthly process of rehabilitation: exploring the wreckage of London with her friend and fellow writer Penelope Fitzgerald, then in her late twenties and half Macaulay's age. They were mesmerised by the wild luxuriance of the bomb sites, whose weeds made them a 'green world' compared to the grey monotony of the rest of London. But Macaulay also felt wistful and ambivalent about this feral regeneration happening in the very heart of London's ancient civilisation (which, like her library, had been burned to a cinder). The irrepressible rosebay and it's blind indifference to human culture made a deep impression: 'Barber's Hall,' she was to write in *The World My Wilderness* a few years later, 'that gaping chasm where fireweed ran into Inigo Jones' court-room.'

The novel, in many ways a distillation of Macaulay's scramblings across the wrecked City, centres on two half-siblings, Barbary and Raoul, aged seventeen and fourteen, respectively, who have run wild with the patriots in southern France, and at the end of the war are sent by their guardians to be civilised in London. They are psychologically unsuited to conventional English society, so instead take to that other *maquis* of the bomb sites, with its raggle-taggle population of squatters, deserters and small-time criminals. And they find the anarchic atmosphere and physical habitat echoing the wild scrublands of southern France, and the Resistance's intrinsic radicalism:

> [They] made their way about the ruined, jungled waste, walking along broken lines of wall, diving into the cellars and caves of the underground city, where opulent merchants had once stored their wine, where gaily tiled rooms opened into one another and burrowed under great eaves of overhanging earth, where fosses and ditches ran, bright with marigolds and choked with thistles, through one-time halls of commerce, and yellow ragwort waved its gaudy banners over the ruin of defeated business men.

Leo Mellor, in his fascinating study of the literature that emerged from the wreckage of 1940s London, has called this 'enfolding verdancy', where nature itself is engaged in 'a regenerative retaking of the city from human destruction', a kind of 'natural amnesia'. But I think it is more than that, closer to a 'natural recollection'; Macaulay was too incisive to invoke some vague green effusion. Just as the lost ancient trades are intoned with the exactitude of names on a war memorial – 'saddlers, merchant tailors, haberdashers, waxhandlers, barbers, brewers, coopers and coachmakers'; and the ruined churches too – 'St Vedast's, St Alban's, St Anne's, St Agnes, St Giles Cripplegate'; so is the City's new weed tenantry: 'a wrecked merchant city, grown over by green and golden fennel and ragwort, coltsfoot, purple loosestrife, rosebay willow herb, bracken, bramble and tall nettles'.

Macaulay and Fitzgerald's safaris into the urban wilderness hadn't just been romantic meanderings. Fitzgerald recalls the 'alarming experiences of scrambling after her . . . and keeping her spare form just in view as she shinned down a crater, or leaned, waving, through the smashed glass of some perilous window'. They were collecting and labelling the bomb-site weeds as they went, and throughout 1948, when she was at work on *Wilderness*, Macaulay was sending cuttings of flowers and shrubs to her friend the writer and amateur botanist Frank Swinnerton for identification. Given the contents of her plant litanies she was also probably familiar with Edward Salisbury's definitive 1945 list of 126 bomb-site flora.

The devotional attention she gives to the ruined City's weeds suggests she saw them as another kind of text, a temporary replacement for those that had been incinerated, and full of thrilling but unsettling reminders of wildness that lay behind civilisation.

> All this scarred and haunted green and stone and brambled wilderness lying under the August sun, a-hum with insects and astir with secret darting, burrowing life, received the returned traveller darting into its dwellings with a wrecked, indifferent calm. Here, its cliffs and chasms and caves seemed to say, is your home; here you belong; you cannot get away, you do not wish to get away, for this is the maquis that lies about the margins of the wrecked world, and here your feet are set; here you find the irremediable barbarism that comes up

from the depths of the earth, and that you have known elsewhere.

～

Macaulay's ambivalent rhapsodies about the 'scarred and haunted . . . wilderness' are an example of the persistence of the cult of the ruin. Ever since the eighteenth century, crumbling buildings – especially those in which rampant weedy growth seems to be aiding or at least capitalising on their decay – have had a double-edged romantic appeal. They may be melancholy records of mortality and impermanence, but they're also parables about arrogance and vanity. Green growth may be destructive of one kind of beauty but replaces it with another. Uvedale Price, one of the doyens of the picturesque movement, described in 1794 how

> Time, the great author of such changes, converts a beautiful object [in this case a piece of Classical architecture] into a picturesque one: First, by means of weather stains, partial incrustations, mosses &c. it at the same time takes off from the uniformity of the surface and the colour; that is, gives a degree of roughness, and variety of tint. Next, the various accidents of weather loosen the stones themselves; they tumble in irregular masses upon what was perhaps smooth turf or pavement, or nicely-trimmed walks and shrubberies – now mixed and overgrown with wild plants and creepers, that crawl among the fallen ruins. Sedums, wall-flowers, and other vegetables that bear drought, find nourishment in the decayed cement

from which the stones have been detached; birds convey their food into chinks, and yew, elder, and other berried plants project from the sides; while the ivy mantles over other parts, and crowns the top.

Price believed that such 'picturesque' processes were partly a demonstration of the 'spirit and animation of nature', and partly an embodiment of the history of a place.

Sixty years later the Sheffield botanist Richard Deakin wrote and illustrated *The Flora of the Colosseum*, an exquisite compendium of the 420 species of wild plant which grew on the 2,000-year-old ruin. There were fifty-six species of grass and forty-one members of the pea family, and some plants so rare in western Europe that they may have arrived as seeds caught in the fur of gladiatorial animals from North Africa. Deakin was most moved by the Christ's thorn (*Paliurus spina-christi*), once worn by the site's ancient martyrs. They were all weeds in the sense of being wild intruders in a cultural artefact, but Deakin saw them as both a testament and a redemption. The Colosseum's flowers 'form a link in the memory, and teach us hopeful and soothing lessons, amid the sadness of bygone ages: and cold indeed must be the heart that does not respond to their silent appeal; for though without speech, they tell us of the regenerating power which animates the dust of mouldering greatness'.

Fifteen years later control of the ruins was handed over to professional archaeologists by Garibaldi's new unifying government, and almost every plant – including those that told more of the Colosseum's history than any of the mute stones – was scoured from the walls.

# II

# *Triffid*

The weed at the end of the world

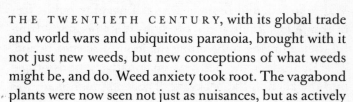

THE TWENTIETH CENTURY, with its global trade and world wars and ubiquitous paranoia, brought with it not just new weeds, but new conceptions of what weeds might be, and do. Weed anxiety took root. The vagabond plants were now seen not just as nuisances, but as actively dangerous. They could invade cities, subvert civilised life, be part of the paraphernalia of modern warfare.

In the summer of 1947 an extraordinary collection of Mediterranean weeds were discovered growing in a bomb crater at Box Hill near Dorking. There were thirty species in all, many of which had never been seen outside gardens in Britain. There were four foxgloves (including the yellow *Digitalis lutea* and rusty-red *D. ferruginea*), elecampane, woad, motherwort, and a scurvy-grass from Turkey. The quantities of some of the species suggested they had already been growing there for three or four years – which meant they had arrived during the war. A rumour began that they'd been dropped with a bomb, either inadvertently attached to it; or, more sinisterly, as some kind of early biological weapon, designed to start an alien weed plague in the Home Counties (a theory which conveniently ignored the fact that any seeds would have almost certainly been incinerated in the explosion). Eventually, a man came forward and confessed to sowing the seeds as an experiment, thinking no one would find them in such an improbable site.

An anxiety about alien plants being used as agents of invading forces wasn't new. During the Second World War a giant puffball was found under an oak tree in Kent, and was suspected of being a new kind of bomb (later it was labelled 'Hitler's Secret Weapon' and put on exhibition to raise funds for the war effort). When the Cold War followed, and brought with it a nagging public fear of communist infiltration, insidious biota became a fruitful theme for the burgeoning medium of science fiction. The Red under the bed was imaginatively transformed into the alien in the backyard. The worst nightmare was that the infiltrator, the shape-shifting outsider, might be mistaken for any ordinary guy. In the film *Invasion of the Body Snatchers* (1956) the amorphous, fleshy templates of the extra-terrestrials hatch from seed pods before taking on the identity of the person nearest to them. The BBC's six-part production of Nigel Kneale's *The Quatermass Experiment* (1953) begins with the first manned space rocket returning to earth with two of its members missing. They appear to have been assimilated by some alien force into the body of the third, which for a while has the superficial appearance of the astronaut who set out on the voyage. During an interview, however, he grabs a cactus and begins to fuse with it. The monstrous hybrid stumbles towards Westminster Abbey, growing all the while, and it soon becomes clear that the cactus has been commandeered in order that its alien parasite can reproduce. As it hangs on the triforium arches above Poet's Corner, root tendrils drooping like some immense mandrake, its spore patches begin to ripen. At the last minute Professor Quatermass pleads with the human psyches locked inside to use their wills to break free. They do, and the vast weed dies. But the penultimate scenes, of the mass of lashing fronds and trunks in the aisles of one of the great bastions of Christendom was, in 1953, the most terrifying thing ever to have been seen on television.

Science fiction's supreme weed saga is, of course, John Wyndham's *The Day of the Triffids*, published in 1951, just before the first H-bomb test in the Pacific. It created an unsurpassable vegetable villain and introduced a new word into the English language. In many ways *Triffids* is a standard post-apocalyptic fable, the tale of a world attempting to survive an abrupt and total social breakdown, although in this case the catastrophe is neither nuclear nor extra-terrestrial. After most of the world's population is made blind by viewing an eerie and mesmerising meteor shower, they become vulnerable to something more earthly: a feral population of mobile, carnivorous – and this was one of Wyndham's many prophetic imaginings – genetically engineered plants.

*The Day of the Triffids* is subtly and intelligently written, and much more than a tale of scary vegetable monsters. The story of the triffids themselves, how they emerged and how they live, is a shrewd insight into the ways that upstart plants invariably become entangled with human needs and cultural preconceptions.

Their background is filled in by the narrator of the book, Bill Masen, once a triffid farmer. The first inkling that such plants existed came when a South American hustler, Umberto Palanguez, attempted to sell the rights for a novel and exceptionally rich vegetable oil to a big international food company. For a large sum of money he promised to provide a supply of seeds. He had seen a picture of the plant they came from, but disclosed little about it, just commenting ominously: 'I do not say there is no sunflower there. I do not say there is no turnip there. I do not say that there is no nettle, or even no orchid there. But I do say if they were all fathers to it they would none of them know their child.' He disappeared with the money, and the next time the oil was heard of was when a Russian gangster arrived at the company offices a few years later, promising to steal a box of triffid seeds from a research station in Kamchatka. Bill already had a hunch that the oil was the product of a plant-breeding programme on the other side of the Iron Curtain.

After a lot of dealing and double-dealing a wooden box was secretly flown out of Russia. But the aircraft vanished over the Pacific, it and its cargo presumably blown to pieces by the security services. And this is where Wyndham first gives a clue of his deep understanding of weed ecology: 'I am sure that when the fragments [of the wreckage] began their long, long fall towards the sea they left behind them something which looked at first like a white vapour. It was not a vapour. It was a cloud of seeds, floating, so infinitely light they were, even in the rarefied air. Millions of gossamer-slung triffid seeds, free now to drift wherever the winds of the world should take them . . .'

It was some time before anyone who knew about the mysterious oil connected it with the curious plants which began springing up in waste corners across the globe: exotic weeds were now a routine fact of everybody's life. Bill Masen's family garden had one 'behind the hedge which screened the rubbish heap'. But as it grew it began to look increasingly odd, disturbingly 'foreign'. It had a straight stem, which sprang up from a woody bole, and three small bare sticks growing beside it. At the top of the stem was a conical funnel or cup, containing a structure like the tightly rolled frond of an emerging fern. Both cup and frond were covered with a sticky substance in which small insects tended to get caught.

'It was', Masen recalls, 'some little while later that the first one picked up its roots and walked' – though it was more like a man stumbling forward on crutches than walking, the two front 'legs' sliding forward, then the whole plant lurching as the rear one drew almost level. A short while later, the young Masen was carefully loosening the roots of his back-garden triffid when he was violently stung and knocked unconscious. The viscous frond at the top of the stem had unfurled and lashed him across the face, leaving a raised red weal.

After the public's initial horror at the fact that these comically perambulating plants could deliver a ferocious sting, interest subsided, until it was realised that the ten-foot-long whiplash of a mature triffid could deliver enough poison to kill an adult human. At first the revelation led to the slaughter of triffids wherever they grew. Later, with their stings docked, and the plants securely fenced in, they became fashionable garden curiosities, vegetable Rottweilers. But in the wilder areas of the planet the triffids' ability to move (and, it turned out, 'lurk') made them a scourge. It also became sickeningly clear that in these remote regions, where large numbers of humans were stung to death, triffids were not just casual snatchers of insects. They were carnivorous. Their stinging tendrils weren't powerful enough to tear flesh from a freshly killed victim, but they could pick shreds from a decomposing body and lift them into the juices in their digestive cups.

By this time the link had been made between these sinister plants and the fabulously nutritious oil that had been smuggled out of Russia a decade before, and triffids had begun to be farmed. They were kept in plantations and planted out in rough rows, with each individual plant chained to a steel stake – a system which also helped keep them safely away from the general public. It was only when triffids were inside these mass corrals that the function of the three little sticks around the stem became clear. From time to time they rattled them against the stalk, sometimes in a kind of tattoo, especially when there were several gathered together. Masen wondered if it this was just a consequence of the wind vibrating the dry sticks when the weather was warm and breezy, or whether it was some pre-pollination love-call, an audible waving of giant stamens. It was his colleague Walter, a more perceptive watcher of triffids than him, who suggested that what they were doing was *talking* to each other.

And that, after the catastrophe of the night of the shooting stars, appears to have been what was happening. As a decimated, hysterical and largely sightless population tries to fend for itself, the triffids break out of their enclosures, track down humans by scent and sound, ambush them, sting them to death and then gorge on their remains. The humans who are blind are defenceless; the few who aren't have to rely on an alertness and a self-restraint that can rarely match the hair-trigger reactions of the monstrous growths waiting patiently by every fence and doorway.

Even as a pure thriller, *The Day of the Triffids* is nightmarish, precisely because of the ordinariness of its details. They have an unnerving modern resonance: the potential killer amusing the family by brightening up the garden rubbish tip; the mass-produced food emanating from a toxic source; the plant that crosses categories from wild oddity to domestic pet to predatory pest – an ancestral trajectory for many weeds. But the triffid was different in that it didn't just move from one category to another; it *fused* them, forming a chimera with both animal and vegetable characteristics (cf. *Quatermass*), and therefore straying deep into the territory of the most troubling human mythology.

Wyndham had created an authentic superweed. The triffid may seem a pure invention, incoherent and implausible. Yet every one of its abilities was modelled on the behaviour of real weeds. Stings, of course, are widespread, carnivorousness less so. But Venus fly-traps are active hunters, snapping their jaws shut over their insect prey, and some tropical pitcher plants can digest victims as big as mice. Plants move, too, not on pseudo-pods (this was Wyndham's least elegant detail) but by rootless slithering, as with the weed dodder, or in furious extensions of their creeping stems, as in species such as Russian vine ('mile a-minute plant'). As for communication by sound, it's not been discovered in the vegetable world yet, but plants unquestionably communicate chemically with each other, and many can register the close presence of other organisms, particularly predators and pollinators, and convey this intelligence to their neighbours. The triffid was an imaginative hybrid of real weeds, and, in the minds of the hyper-anxious, not beyond the future bounds of biotechnology. As Umberto Palanguez says of his first glimpse: 'I do not say that there is no sunflower . . . no nettle, or even orchid there . . .'

~

In 1970 a real triffid entered the public's imagination. That summer, large numbers of children began turning up at hospital casualty departments with unusual red weals and circular blisters on their lips, hands and eyes. The blisters seemed more prevalent during sunny spells, and were apt to leave long-lasting and angry scars. The source of the problem was traced to a tall umbelliferous plant called giant hogweed. It didn't lash out at humans like its fictional counterpart. But it did contain photo-sensitive chemicals called furocoumarins that were activated in sunlight, and were causing burns on children who had been hacking down its thick hollow stems and using them as blowpipes and telescopes. The revelation that this weed was an immigrant from the other side of the Iron Curtain, and a touch too smart for its own good, made the story hot news. The popular press immediately dubbed it 'The Triffid', and published full-page photos of its speckled fifteen-feet-tall stems and cartwheel-sized flowerheads. They also published tips about how to eradicate this Ruskie menace: poison, flame-throwers, excavators, digging out the roots and filling the hole with turpentine, even dynamiting. None had much effect, and giant hogweed continued its long march across Britain.

What is curious is how long it took for the plant, and its effects, to be noticed. Giant hogweed was first introduced to Britain from the Caucasus in the early nineteenth century, and was widely admired as the temperate zone's

answer to the titanic plants being brought back trium-
phantly from the tropics. One of the arbiters of Victorian
gardening taste, John Loudon, was lavish in his praise of
'*Heracleum asperum* . . . the Siberian giant parsnep' in the
*Gardener's Magazine* in 1836:

> The magnificent umbelliferous plant, when grown
> in good soil, will attain a height of upwards of 12ft.
> Even in our crowded garden in Bayswater, it last
> year was 12ft when it came into flower . . . Its seeds
> are now (29 July) ripe; and we intend to distribute
> them to our friends: not because the plant is use-
> ful, for we do not know any use to which it can
> be applied; but because it is extremely interesting
> from the rapidity of its growth, and the great size
> which it attains in five months . . . We do not know
> of a more suitable plant for the retired corner of
> a churchyard, or for a glade in a wood; and we
> have, accordingly, given one friend, who is making
> a tour in the north of England and Ireland, and
> another, who is gone to Norway, seeds for deposit-
> ing in proper places.

(Whether this Scandinavian Johnny Hogseed was
responsible for the now extensive Norwegian populations
of the plant is not known. But giant hogweed – known in
the north of Norway as 'Tromso palm' – is regarded with
more affection there than it is in Britain, and appears as a
feature on local tourist postcards.)

By 1849 seeds were being sold commercially by Hardy
and Sons, of Maldon, as *Heracleum giganteum*, 'One of the
most magnificent Plants in the World'. The advertisements
suggested that, for just a few pennies' worth of seed, the

ordinary gardener could grow, out of doors, plants every bit as majestic as the exotic marvels being raised in the hothouses of the rich. In 1870 it received the endorsement of William Robinson, the champion of informal gardens, though he did warn of the dangers of it becoming a 'giant weed'.

In the early 1900s giant hogweed (now carrying a more appropriately extravagant Latin name, *H. mantegazzianum*, in honour of its Italian discoverer), began breaking out of the elegant water gardens and estate woodlands where it had been sown. A collection in Buckingham Palace Gardens edged into the Royal Parks, and thence into the west London canal system. The floating seeds spread freely along watercourses, and many early colonies, especially in Scotland, were traceable to big country houses with riverside gardens. A map of the plant's distribution in 1963, seven years before it hit the headlines, shows populations scattered the length and breadth of Britain's river systems. It did no good to the Hog's reputation that it seemed (like the pheasant) to be an invader from the milieu of another class.

For all this time there seems to have been no suspicion of the weed's photo-dermatitic effects. But when these became news in the 1970s the situation changed dramatically. Giant hogweed was included in an appendix to the Wildlife and Countryside Act of 1981, and it became an offence to plant it or knowingly tolerate its growth in the wild. Some local authorities went beyond the letter of the law, and requested that householders destroy visible specimens inside their own gardens. I saw a roadside specimen in Middlesex fenced off by 'Keep Out' tape, as if it were a crime scene.

There was another reason for its inclusion as an

undesirable alien in the 1981 Act. It had become quite invasive, especially near rivers and wetlands, and there was concern amongst conservationists that its immense jagged leaves were blanketing out native plant species. In East Anglia, for example, it now grows along rivers in the Norfolk Broads, and in an orchid-rich fen near the sea in Aldeburgh. These colonies unquestionably look 'out of place' amongst the native vegetation – though my own impression is that they're not causing much harm to the plants that grow in their light shade.

Giant hogweed – despite its alien origins, the biggest and most architecturally dramatic non-arboreal plant to grow wild in Britain – was always destined to spark fiercely contrary feelings. For Hog aficionados, there are landmark colonies across Britain, places worth visiting just to see the massed white coronas: by the sides of the Toll Bridge on the outskirts of Nottingham, for instance, around the Launceston Recycling Centre in Cornwall and along the River Usk near Abergavenny. One of the oldest and most famous colonies is in the damp wasteland outside the Hoover Factory in west London, where the leaves, a passable if gross modern variation on the Classical acanthus motif, perfectly set off the listed Art Deco buildings. (I first saw this patch at a poignantly theatrical moment, when roadworks on the adjacent A4 were adorned with signs reading 'Heavy Plant Crossing'.) The biggest and most awesome hogweed forest is probably on the banks of the Clyde in Glasgow. Jim Dickson, Senior Lecturer in Botany at Glasgow University in the 1990s, has described the immense sweep blanketing the riverbanks and waste ground downstream from Kelvinbridge as 'one of the most remarkable natural history sights of the Glasgow area'.

People in Scotland, because they have lived with the plant for rather longer than other Britons, seem to favour an approach based on cautious respect rather than knee-jerk extermination. Many parents, rather than demonising the plant, teach their children to recognise it and keep their distance. In Glasgow itself, the high ceilings of the older terraced houses have given householders the opportunity to use dried Hog stems in flower arrangements.

Inspired by this idea, we brought some sprays inside our own house in 2006. A couple of specimens were growing by a farmer's rubbish pile in the neighbouring field, and in the late autumn, carefully begloved, we cut off the dried stalks and stole them home to use in place of a Christmas tree. The flower-stalks and umbels (still carrying their seeds) were surprisingly strong and rigid, and we were able to festoon them with baubles and stars and a few model owls. In the spring we manoeuvred the stalks upstairs into a big vase, while outside in the garden their shape became the inspiration for a bird feeder made of umbelliferous starbursts of scrapyard iron mounted on top of a sycamore log.

But we'd reckoned without the indomitable mobility of weed seed. In the autumn of 2008 a mysterious rosette of serrated leaves appeared in the gravel two feet from our front door. I couldn't make any sense of it to start with, but by the following spring it was all too obviously a young giant hogweed, and by June it was in full flower. It eventually reached up to the foot of the thatch, and we were constantly needing to trim the leaves back and truss up the sprays to make the doorway safe for visitors, especially our beloved post lady. But unlike our unruly grass verge, no neighbour ever reported it to the Council, and the giant hogweed – for all its majesty, a weed with a lifespan of just twenty months – never appeared again.

~

Weeds have had more constructive roles in the fiction of apocalypse and social breakdown. Richard Jefferies' fantasy, *After London, or Wild England* (1885), imagines the sudden collapse of English civilisation after a catastrophe. (This isn't specified, but seems to have been a flood which submerges London and converts the Thames Valley into a vast lake.) The first part, 'The Relapse into Barbarism', is a meticulous naturalist's account of the spontaneous return of 'The Great Forest'. It is spelt out step by ecological step.

All agricultural land has been abandoned, and during the first spring 'after London ended' arable fields are overrun with couch grass. By the summer, former roads and tracks are thinly covered with grass that has spread out from the margins. The following summer, after a rain-sodden winter, the weeds have their moment. The bird-sown wheat and barley that pushes through the uncut and rain-battered straw is accompanied by thickets of docks, thistles, ox-eye daisies, charlock and nettles. Each year a few crop plants reappear, but they're a declining presence, and soon smothered by the big perennial weeds advancing into the fields.

Jefferies understood the process of plant succession exactly. He would have witnessed it happening in the abandoned fields of impoverished farmers in his native Wiltshire. He knew that brambles would be the next species to advance into the fields from the hedgerows, followed by the wild roses. Soon the hedges would be three or four times their original breadth, and after twenty years or so would meet in the centre of the biggest fields. In the damper places 'beside the flags and reeds, vast quantities of the tallest cow-parsnip or "gicks" rose five or six feet high, and the willow-herb with its stout stem, almost as woody as a shrub, filled every approach'. By the thirtieth year, with all ditches choked by dead leaves and fallen branches, the overflowing water had turned most low-lying fields into marshes. As for the drier ground, it had become a natural forest of oak, ash and thorn, 'where there was not one single open space, the hills only excepted, where a man could walk unless he followed the tracks of wild creatures or cut himself a path'. The movement was inexorable, from pioneering weeds to impenetrable high forest in just three decades.

Twenty years later, in the effeteness of the Edwardian era, Kenneth Grahame suggests that a similar process created the Wild Wood in *Wind in the Willows*. Badger explains how the wood had 'planted itself and grown up' on the site of a collapsed city. (Grahame himself was a denizen of the City, simultaneously Secretary of the Bank of England and a paid-up member of the Pagan Society.) There are hints in the text that Badger's sett itself has been built in the derelict cellars and tunnels of the vanished metropolis. He explains to Mole the slow regeneration of the forest: 'It was all down, down, down gradually – ruin and disappearance. Then it was all up, up, up gradually, as seeds grew to saplings, and saplings to forest trees, and bramble and fern came creeping in to help.'

Periods of social and political tension always bring prophecies of a showdown between civilisation and 'the

weeds and the wilderness', with nature usually given odds-
on to win. In the strained debate about European agricultural surpluses in the late 1980s, the *Observer*'s trenchant
columnist Neal Ascherson painted a dystopian vision of
what would happen to agricultural land left to become fallow, or deliberately 'set aside'. His prophesy echoed that
of Richard Jefferies, but had a barbed modern twist in the
tail:

> Left to themselves, the fields would fur over with
> weeds, waist-high and then head-high. Bushes
> would be followed by small trees, and eventually –
> in most of lowland Britain – by dense and scrubby
> secondary forest. Much of this land would revert to
> waterlogged swamp, as field drainage broke down.
> It would be good for birds, but also good for rats,
> mosquitoes and accumulations of weed pollen to
> make the nation sneeze and to smother its gar-
> dens. In the dimness of the tangled undergrowth
> there would lurk, like the debris of a forgotten bat-
> tle, millions of abandoned cars, refrigerators and –
> especially – agricultural machines.

Vegetable trash was destined to attract, by a kind of
sympathetic magic, human rubbish. Weeds were not only
a consequence of dereliction, they were a cause of it, and
then, paradoxically, the means by which it was obliterated.
Will Self's sprawling fantasy *The Book of Dave* (2006) is
one of the first futuristic novels to take climate change as
its apocalyptic premise. The story is set some 500 years
in the future. The oil has run out, rising sea levels have
(as in *After London*) flooded the Thames Basin, and all
the low-lying ground around the river's tributaries. The

remnants of civilised society have mostly retreated to hilly areas like 'Cot' and 'Chil'. All that's left of east London is an island called Ham, where the lives of a depleted tribe of hunter-scavengers are ruled by the Book of Dave, the unearthed jottings of a demented twenty-first-century cabbie. A remarkable feature of the book (and one that makes its long glossary essential) is the language Self invents for this post-urban society, a mash of pidgin Cockney and mutated brand-names from the folk-memory. ('Starbuck' is breakfast, for example, and 'A2Z' a map.) One of the Ham tribe's annual rituals are raids upon the seabird colonies that have colonised the remains of the City of London's skyscrapers ('stacks'), to gather 'oilgulls' (fulmars), 'prettybeaks' (puffins) and 'blackwings' (gannets). There's a local legend that if a climber falls from a stack he'll be rescued by a formation of seabirds called a 'choppa'.

Self sets this picaresque saga amidst a vegetation which might plausibly have developed from the contemporary weeds of London's East End, or of the ramshackle landscapes of the Essex estuaries. 'Rhodies' (rhododendrons) cluster most thickly round the edges of what were once big houses, and 'pricklebush' (furze), is intermingled with it. But the wastelands and long-derelict industrial sites that slope down to the water's edge are covered with an authentic urban scrubland: 'fireweed' (Self uses the traditional vernacular for rosebay), 'blisterweed' (a sharp coining for giant hogweed that could well catch on), 'burgerparsely' for cow parsley and 'buddyspike' for buddleia.

∾

In 2007 Alan Weisman's astonishing non-fiction book *The*

*World Without Us* put all these well-informed but imaginary predictions into perspective. The premise of the book, its literary conceit, is outlandish but fruitful. Weisman has the entire human population of the earth disappear 'in the twinkling of an eye', as in the Religious Right's notion of 'The Rapture'. He then tries to map out what would happen to the planet without our daily, unrelenting attacks on nature, based on the hard evidence of what has happened in real-life cases of human abandonment.

He starts with the evidence of a single house. The speed at which plants help to deconstruct an abandoned building is startling. In the first winter, rain begins to rust exposed nails, and penetrate the holes around them. Mould and fungi penetrate the damp wood and break it down. The same thing happens to the floors. As the wood starts to crack and decay, roots from the larger weeds and trees outside begin to invade the house, penetrating the crumbling timbers and breaking them up still further. With no heating, pipes burst in the first freeze-up, creating small flood patches which are rapidly colonised by duckweeds, nettles and young tree seedlings. The cellars, probably now open to the sky, fill up with debris, and brambles start ambushing remaining pipes. If there's a swimming pool, it's now an indoor water-feature, filled either with the offspring of exotic house-plants, or with banished weeds that were hovering in the margins of the property, waiting for the chance to retake their territory. Within fifty years, all that is left of the house is a vague swelling in the earth, a post-industrial tumulus, specked with shards of non-degradable plastic and porcelain tiles, and covered with a mixture of trees and naturalised garden plants.

In New York, it's already clear that just a few months

of neglect by city maintenance teams would lead to the streets becoming a burgeoning forest of Chinese tree-of-heaven seedlings. The winged seeds would lodge in pavement cracks and subway tunnels. They germinate and grow fast, the tree's name being a reference not to its paradisiacal properties (its flowers smell rather disagreeable) but to the speed at which its saplings rocket skywards. A few months later the spreading leaf clusters will be poking through sidewalk grates, and the powerful root systems (which also send up suckers), will be heaving up pavement slabs and cracking open sewers. Within a decade the trees could be more than thirty feet tall. And as soil long trapped under the pavement becomes exposed to sun, rain and sewage nutrients, ground-weed species will jump in and form an understorey beneath the burgeoning saplings.

This is precisely what happened to an abandoned section of the New York Central Railroad in Manhattan. The track was closed in 1980, and the trees-of-heaven invaded immediately, soon joined by lamb's-ears and golden rod. In some places, the track emerges from the second-storey warehouses it once serviced, and now carries ribbons of irises, evening primrose, Michaelmas daisy and Queen Anne's lace (wild carrot) way above the ground. Many of these species, like the city's human inhabitants, are immigrants from Europe and the Far East, and maybe it is no surprise that, in Weisman's words, 'So many New Yorkers, glancing down from windows in Chelsea's art district, were moved by the sight of this untended, flowering green ribbon, prophetically and swiftly laying claim to a dead slice of their city, that it was dubbed the High Line, and officially designated a park.'

Modern Detroit is the High Line writ large. In the 1920s it was one of the richest cities in the world, grown fat on the products of Ford and General Motors. But the Motor City was as vulnerable as all monocultures. When the oil crisis began in the 1980s the automobile industry began to decline, and eventually left Detroit altogether. The city, with no other large source of income, began physically to collapse. Derelict factories, and the abandoned homes of the people who had once worked in them, began to be reclaimed by nature. Prairie weeds have colonised the parking lots and deserted freeways. Wild vines like kudzu are scaling walls, and trees-of-heaven thirty feet tall have sprung up on factory roofs. There are now 66,000 vacant lots, and 40 of the 139 square miles of the inner city have been taken over by wild vegetation, which is in the active process of demolishing what was once the fourth largest city in the United States.

But the response of the human inhabitants has been unexpected. There's been no horrified backlash against this invasion, no attempt to scapegoat nature for what is at root, the result of economic and political stupidity. Instead, the weeds are being read as a parable, a lesson that a monolithic, oil-based urban culture is unsustainable in the twenty-first century, and that there might be other, more ecologically gentle ways of living in cities. Families too poor to buy fresh food are starting neighbourhood organic farms on the sites of demolished local blocks. Young people from all over America – musicians, Green activists, social pioneers – are flooding into the abandoned areas, keen to experiment with new patterns of urban living which accept nature – including its weedy frontiersmen – rather than attempting to drive it out. As

Julien Temple, director of the remarkable TV documentary *Requiem for Detroit*, has written: 'amid the ruins of the
Motor City it is possible to find a first pioneer's map to the
post-industrial future that awaits us all'.

~

The literature of the apocalypse presents weeds in ambig-
uous, not to say contradictory roles. They may be one of
the agents that bring down civilisation, but they may also
be the living pioneers that begin to rebuild it. In both
scenarios, what drives the action is an outlaw, an organism
from another place or culture, beyond the compromised
and collapsing world of hapless humans.

In the real world, there is abundant evidence of both
scenarios. In the tropics destruction is the prevailing
mode. Fast-growing plants translocated from their native
lands to be cultivated as, say, forage grasses or fast-growing
timber crops, have turned into weeds that are ravaging
entire ecosystems. The ecologist Jonathan Silvertown has
described them graphically as 'demons in Eden'. Florida
– hot, damp, and in a state of constant upheaval from
development – has been especially badly hit. In the 1930s
Australian paperbark trees were introduced to the Everglades in the hope that they would dry out the marshes
sufficiently to grow crops and condominiums. The seeds
were sown from aircraft directly into the pristine swamp-
land, where they sucked out five times more water from the
ground than the native species. In their native Australia
paperbarks are attacked by constellations of insects, but
in Florida nothing eats them, and the trees have grown
furiously. Each tree begins producing wind-borne seeds at

just two years old, and releases 20 million of them a year. At the height of its invasion, paperbark occupied 1,000 square miles of southern Florida, at a density in places of nearly 8 million trees per square mile. Elsewhere in Florida, Brazilian pepper, introduced from South America as a garden shrub, has escaped to form immense stands that smother all native plants, partly because, in addition to producing huge quantities of well-dispersed seed, it's an accomplished climber, and seems to be toxic to many of the plants it comes in contact with.

The remaining areas of open water in the Everglades are being carpeted by alien aquatic weeds, some of them throw-outs from garden ponds and aquaria. The most troublesome – and a weed that is now choking wetlands and river systems in fifty-six countries in the tropics and sub-tropics – is water hyacinth, a native of Central and South America. It's an attractive plant with spikes of purplish flowers and clusters of billowing glossy leaves that are raised above the water to act as sails, and it's easy to see why it became popular as an ornamental. In its native wetlands it behaves itself, but elsewhere it can double its population in fourteen days. Each individual plant is free-floating and supported by air-filled bladders, and with the help of its leaf-sails can move easily about on the water, producing new offspring both by seed and by budding. It can be controlled with herbicides but with inevitable collateral damage to native aquatic life.

A little further north is perhaps America's worst demon, the kudzu vine introduced from South-East Asia in the 1870s. As so often, the intentions behind its introduction were honourable. The 1876 Centennial Exposition in Philadelphia contained a Japanese garden full of that country's native plants, kudzu included. It was a popular exhibit and American gardeners began planting kudzu as an ornamental. Its initial spread from cultivation was comparatively slow, but in the 1920s a Florida nursery noted that cattle were browsing on kudzu, and began promoting it as a forage crop. Ten years later the Soil Conservation Service began planting out the vine to help control 'Dust Bowl' soil erosion – the result, ironically, of an earlier episode of short-term agricultural thinking. By the 1940s the US government was paying farmers up to eight dollars per acre of planted kudzu. Within a few years its advance was unstoppable.

To describe kudzu as an aggressive climber makes it sound like a rather pushy rose, a Rambling Rector, say, not a demon's disciple. But in the height of the growing season kudzu can grow a foot in twelve hours, and a joke in the southern states is that you must close your windows at night to stop the vine getting in. Abandoned buildings can quickly disappear under a blanket of the weed, as can whole stands of native forest. It can reach a height of ninety feet, by which time the trees supporting it are dying from lack of light. Nothing so far seems able to oust it. Contemptuous of most hallowed ecological principles, it has dug itself in and become a seemingly stable climax vegetation.

In the American South kudzu now covers 2 million acres of forest land, and has been officially outlawed by the US Department of Agriculture. But statistics don't give any idea of the experience of living with this remorselessly advancing cloak. A kudzued landscape is eerily beautiful, if you don't stop to think about what it has buried. It has a primordial aura, as if an ancient city had been overwhelmed by the jungle. The trees look as if they have been petrified by green lava, or a monochrome coral, or are the seaweed-enfolded relics of a wrecked ship. Some Americans have found comfort in its irrepressible greenery. The writer Francis Lam, fleeing to Alabama to escape Hurricane Gustav, was transfixed by it: 'There's kudzu growing on power lines, kudzu growing on trees, kudzu growing on buildings, kudzu growing on kudzu growing on itself. Every few minutes we'd come across another field of lush vines and every time the feeling was like waking up after a snowstorm and seeing the whole world's corners rounded off. The sight was stunning and I was quietly grateful for it, a respite from the tension we felt while evacuating from the hurricane.' In the South, despite its rampaging spread, kudzu is still used for controlling erosion, and the stems harvested for basket weaving. Regular grazing, for which it was originally promoted, is regarded as the most economical method of limiting its spread.

A weed of such hyper-natural expansiveness and mysterious aura was bound to generate – as alien weeds have in the past – myths about its origins. The most extravagant (on the US website 'Mindspring') involves a conspiracy theory so outrageous that it may well be a spoof, but which nonetheless perfectly encapsulates the paranoia felt about botanical invaders.

Mindspring readily acknowledges kudzu's Asian origins but insists that its introduction to America was part of a sinister plot by 'Japanese secret operatives' to subvert the US economy. They targeted the Forest Service, which was supposedly already looking for solutions to eroded soils back in the 1870s. And here a familiar device in the mythology of alien plants enters the plot: an ingenious and discreet conveyance in which the plant is able to travel across continents. Kudzu seeds were sent to America anonymously, in an envelope with no return address. They were planted out by a grateful Forest Service and the green menace began to spread. Thirty years later, more envelopes arrived, this time containing even more invasive varieties of kudzu, developed in secret Japanese laboratories. By the 1940s the authorities at last realised that the whole national forest was being devastated, and an eradication programme was developed involving the combined manpower of the National Guard and military reserve forces. The Japanese knew they must do something drastic to derail the kudzu pogrom, create a distraction so provocative that it would divert all that military effort. No prizes for the answer. They bombed Pearl Harbor. The United States, preoccupied with the war effort against the most conspicuous Japanese threats, ignored the insidious spread of their real secret weapon. Ever since, kudzu has continued to advance across the United States while the Japanese buy up real estate in the cities. 'Maybe you or I won't be here to see it,' Mindspring concludes, 'maybe our children's children will be the ones to shriek the final silent scream of Kudzu-muffled terror as the last little piece of blue sky is covered over by the unstoppable wave of fuzzy green growth.'

This theory might be implausibly paranoid, but bears comparison in terms of both language and content with strictly scientific accounts of the alien plant invasion of Australia. Just why the southern hemisphere and Australia in particular has suffered so acutely is complicated. There are plenty of contributing factors: the isolation of

the continent, and a flora that has few genetic or biochem-
ical connections with plants in other parts of the world; the
climate, hot and wet by turns; and the thin, nutrient-poor
soils, which have had no ancient history of hoofed animals
to scuff them up and encourage the evolution of plants
resistant to disturbance (i.e. native weeds). But suffer
badly it has, enough for the crusading Australian biologist
and writer Tim Low to write an entire book – *Feral Future*
(1999) – about the catastrophic effect exotic invaders have
had on Australia's indigenous organisms. It is an angry
book. Low spells out that the country now has more than
2,500 species of rampaging alien weeds, costing the economy A$4 billion per year, and he talks of their entry into
Australia's ancient cultures as both an example and a consequence of globalisation. His language can be uncomfortable for temperate-blooded Europeans who can have
no conception of the scale of the botanical invasions he is
describing. He talks as if he blames the invading plants
themselves. The invaders 'steal' into our forests, 'foul' our
rivers. Weeds 'fester'. St John's Wort, source of an effective anti-depressant, is described as 'malevolent', mimosa
as the 'repugnant claimer of 30,000 acres of wetlands'.
Nor is it easy for Europeans, with their long experience
of ancestral and relatively harmonious blendings of natural and cultural vegetation, to sympathise with statements
such as 'Olive trees round Adelaide have converted whole
hillsides into vast gloomy thickets'; or that more 'gloomy
thickets' of Europe's beloved blackberry, deliberately
scattered about by early English settlers, could now be
regarded as one of Australia's most 'noxious' weeds, costing A$40 million a year to control. It needs a determined
effort of cultural empathy to comprehend a weed threat

when the plant in the wrong place is adrift by 12,000 miles. Low's emotive vocabulary doesn't help his case. When he says that white horehound (introduced from Britain in the mid-nineteenth century as a medicinal herb) now 'infests' about 25,000 square miles of Victoria alone, does he mean it is scattered over that area or in exclusive occupation? The difference matters.

But we're in no position to quibble. Most of the early alien plant invaders of Australia, as of the United States, came from Britain. The painter Marianne North, travelling in Tasmania in 1881, complained about the thistles, docks and dandelions: 'The country was not in the least attractive to me; it was far too English.' But a telling early example of how comprehensively the biology of this isolated island continent could be affected by unfamiliar organisms was sparked off by a weed from South Africa. Capeweed, a yellow daisy from the Cape of Good Hope, arrived in the mid-nineteenth century and set in motion ecological ripples that continue to this day. Within twenty years its pale lemon flowers were 'covering whole hillsides and every vacant spot'. 'It grows in knee-deep masses by the wayside', reported the botanist J. E. Tenison-Woods in some alarm. Then, in 1889, the larvae of a native butterfly, the Australian painted lady (*Vanessa kershawi*) discovered they could eat the leaves. The daisy was so plentiful that the butterfly went through a population explosion, darkening the skies across Victoria. There were newspaper reports of trains coming to a halt in one railway tunnel because the masses of crushed butterfly bodies had lubricated the wheels to such an extent they could not grip the rails.

And as in the United States, it was the combination of

new plants and new animals that caused the most trouble. The tough pasture grass, buffel, was probably first introduced to Australia from Africa with the camel, but it provided such good grazing that it was soon being deliberately sown. Yet it was only the immense increase in the number of sheep that caused its population to explode. Australian vegetation, quite unused to hooved and rapacious stock, was no match for plants that had evolved to be tolerant of grazing, and become quick growing, aggressive seeders in response. Another pasture plant – an Asian pampas grass – remained a sleeper for almost a century because all the plants were white-plumed females, unable to set seed. Then, in the 1970s, someone introduced a new pastel form that happened to be a pollen-bearing hermaphrodite. This crossbred with the females and yet another new Antipodean weed was born.

Almost any alien plant seems capable of becoming a weed in Australia: daffodil, sweet pea, lavender, peach, olive, willow, grape, fig, carrot, sweet-briar, watercress, cashew, peppermint . . . the list increases by dozens every year. Only in very few cases has the invader been successfully and sustainably controlled. During the 1920s, prickly-pear cactus (from sub-tropical America) was ranked as probably the most widespread weed on earth. In Queensland and New South Wales, Low reports, 25 million acres were 'infested' with it. Then, in what has become almost a biblical parable amongst believers in the organic control of pest species, a moth caterpillar from Argentina, *Cactoblastis cactorum*, was brought to Australia in 1925 by a concerned entomologist and released. It immediately began to munch its way through the prickly pear population, and 'by the end of the second year, the countryside for mile after mile was covered inches deep with rotting pear plants and a slimy jelly-like substance'.

The defeat of the cactus still stands as an outstanding example of biological control. But it has proved a double-edged solution. *Cactoblastis* has spread across the globe, and begun to destroy populations of wild, native cactuses. Organic herbicides can become as big a problem as the weeds they control.

Australia has problems with alien invaders that are almost too complex to think about. But there is no realistic hope of returning it or any other country to some idealised and stable 'natural' state. Ecosystems are – and must be if they are to remain resilient – malleable, adapting to climate change and extinction. Nor can alien invaders be wiped off the earth with a fit of pique or burst of herbicide. The best we can do is try to find ways of incorporating into our lives and ecosystems those we already have, and attempt to prevent the arrival of unhelpful newcomers.

Low quotes an intriguing experiment by the botanist Jamie Kirkpatrick: 'Only 1.1 to 1.4 days of normal individual human urine output will fertilise a square yard of soil sufficiently to raise it from being able to support only native heath to levels suitable for most crops and most weeds.' It's worth bearing in mind that the first alien creatures to arrive in Australia, the first pissers on its indigenous virgin soil, and the initiators of the whole business of organism translocation, were not European colonialists but the Melanesian explorers who set foot in a human-free Australia 45,000 years ago.

~

Alien plants have created nothing like the problems in
Britain that they have in Australia. Our climate is cool,
and unconducive to most of the sub-tropical invaders that
are creating havoc in warmer regions. Our native vegetation is used to being grazed, cut back, roughed up, and
has developed more resistance to penetration than pristine
ecosystems. There are, as we'll see, the knotty problems of
Japweed and Indian balsam, but most of our new weeds
are environmental nuisances rather than ecosystem wreck-
ers like kudzu and buffel. A single red-hot poker flaring
amidst the austere poles of a conifer plantation, a cannabis
prong lurking under the churchwarden's bird feeder, are
no more than botanical puns, mischievous instances of the
enterprise of plants rather than ominous portents of alien
invasion. At least that's the way I see them. I'm afraid I've
never lost the thrill I felt back in the 1960s at discover-
ing that riotous community of cosmopolitan squatters in
the Middlesex wastelands, or my amazement at how they
arrived there. I've nosed out odd plants in odd places ever
since, and – though this may sound like rationalisation for
what is little more than botanical slumming – feel that I've
picked up some clues as to why aliens are not the threat
here they are in the hotter parts of the world.

Those lunchtime safaris from my Penguin offices led
to a book called *The Unofficial Countryside*, and then a television film. These weren't concerned just with eccentric
weeds, but with the totality of urban nature, the heartening incongruity of the wild co-existing with that crowning
achievement of civilisation, the city. They were about kes-
trels nesting in tower-block window-boxes, about foxes
strolling up Whitehall, about the tropical clothes moths
that appeared at Buckingham Palace after a garden party

for Commonwealth Prime Ministers. But the city's feral vegetation was always there in the background, a frisson of immanent luxuriance.

Rosebay was, of course, a constant backdrop during the summer, especially in the tarry wastes of the derelict gasworks at Beckton, where it framed our sequences of black redstarts. But often the weeds themselves were the story-line. On the railway embankments near Willesden Junction we filmed an extraordinary wild garden, drifts of luxurious fruit and vegetables, under cultivation decades earlier but now forgotten and – very much in the wrong place – growing on as weeds. There were asparagus clumps six feet across, entanglements of loganberries, fruiting blackcurrant bushes. Their history was intriguing. During the Second World War, when the whole population was being urged to 'Dig for Victory', the householders whose properties backed on to the embankments decided that railway land shouldn't be exempt from the war effort and extended their personal vegetable plots up to the edge of the line. These makeshift allotments were abandoned after the war, but the crops (the perennials, at least) survived.

Provenance, persistence and peculiarity are the necessary ingredients for a weed to enter the connoisseur's list. At the back of the Ford motor works at Dagenham in those days there was a considerable colony of tumbleweed, the plant whose balls of dried stalks bowling about the badlands is an essential part of the ambience of Western films. (Tumbleweed's nomadic habit is, needless to say, a classic weed strategy, evolved for the vast expanses of the desert. The parent plants dry off after flowering, are uprooted by the wind and blow about the landscape, scattering seeds as they go.) The irony is that this is one of the choicer examples of cinematic anachronism. Tumbleweed – alias Russian thistle, *Salsola kali*, subspecies *ruthenica* – is a native of the arid areas of eastern Europe and Asia, and only arrived as a weed in the United States (mixed up with flax seed brought by Ukrainian immigrants) in the late 1870s, some while after the pioneering heydays portrayed in classic Westerns.

Russian thistle was first recorded in Britain in 1875, in a garden on the outskirts of Oxford. It had probably been brought in with wool-waste manure. The Dagenham colony was spotted in the 1930s, its progenitors having presumably hitched in with the imports from Ford America. What clinched their successful germination and spread was finding a habitat remarkably like the dry plains of the Russian steppes, or the American West. At the back of the works was a large area where the ashy waste from the foundry was dumped. It was a warm, dry, shifting substrate, an industrial desert, and the tumbleweeds took to it like schoolkids to sand dunes.

By 1934 the colony consisted of hundreds of plants. When we came to film them in 1974 much of the waste tip had been levelled and turned into a parking area for Ford models hot off the production line. But a patch of the weeds remained, and there was a good wind blowing that day, a desert breeze, so we got some unique footage of Essex tumbleweeds attempting to fulfil their genetic destiny, and blundering into the security fences round the new Cortinas.

By then I had a taste for aliens, partly because of the often extraordinary means by which they'd arrived, but also because of their disregard for the proper botanical order of things, their sheer opportunism. I loved the way that available spaces were filled indiscriminately by weeds which were either deeply improbable or historically inappropriate, in nature's proverbial abhorrence of a vacuum. I fear my appetite was a shade forensic at times. The most extreme example I came across of a plant in the wrong place was in a medical journal article about an alfalfa seedling that had sprouted in the moist warmth of a patient's eyelid. Much less discomfiting was finding a deadly nightshade growing directly out of a tomb in St Cross churchyard in Oxford. I like to think it may have had a similar origin to the notorious 'Atheists Fig' that sprang out of a grave in a Watford cemetery in 1913. (The legend was that a local unbeliever asked for a fig to be placed in his hand in the coffin, saying that if there was life beyond the grave he would make it sprout, though the tree more probably originated from a last snack taken by the hapless occupant.)

Occasionally I went on 'Alien Hunts' organised by the Botanical Society of the British Isles. This was a euphemism for coach tours round east London refuse tips, which in those pre-recycling days were sites for the indiscriminate dumping of every kind of rubbish from abattoir waste to plastic toys. In early autumn we would saunter through a landscape of freshly dumped cows' intestines, windblown paper, broken glass and immeasurable piles of stinking kitchen refuse. The whole landscape lay under an acrid mist from smouldering rubber fires and steaming organic refuse. Lorries shuttled to and fro, bringing new loads of garbage, while bulldozers flattened it into a kind of compacted compost. I doubt that any natural habitat on earth has quite the unceasing frenzy of disturbance of the old-style tip, nor such a concentrated input of exotic seeds. It was a forcing bed for weeds.

And they came up in their thousands, a bizarre mixture of food plants and herbaceous ornamentals and southern hemisphere stowaways lodged in imported pot-plants. They sprang from thrown-away salads and the refuse from oriental restaurants and the thousands of tons of birdseed imported every year. They rooted from hedge clippings and fruits embedded in mud-slicked tyres or import packaging. Every viable seed, or fragment of root or stalk, stood a chance of being nurtured into a viable plant by this warm garbage. We found buckwheat, canary-grass, coriander, cucumbers, cumin, dahlias, a single plant of that old cornfield plague, darnel (probably an impurity in imported barley), dill, fennel, fenugreek, gourds, henbane, iris, love-lies-bleeding, marigold, marijuana, nightshades (five species), potatoes, Russian vine, shoo-fly plant, sunflowers, tomatoes (fruiting), watermelons and *Xanthium spinosum*, a comic-strip cocklebur from South America, whose fleece-entangled fruits arrive at refuse tips with wool waste.

When an especially choice species like this was found, the expedition leader would blow a whistle and the members of the party would leave their private meanderings and cluster round the plant. Photos would be taken, and there would be an earnest debate about the finer points of identification. If this was in doubt, a decision would be taken about who should 'carry on' the plant, until it revealed its identity. This involved digging the plant up (it would be smothered by new rubbish in a few weeks anyway), inserting it in a damp polythene bag and taking it back to a home greenhouse or botanic garden for nursing through to the revelatory moment of flowering or seeding. On one lap between tips, the warmth inside the coach made the shoo-flies (closed up when they were harvested) open their glorious blue flowers *inside* the polythene bags. Thinking of the oddity of this plant's, name, I remembered Posy Simmons's mischievous cartoon linking the folksy English names of plants with the twentieth century's new breed of rural detritus. Her caption seemed peculiarly apt in these surroundings. 'The lane banks are gay with DROPFOIL, GUMBANE and YELLOW CORN COCKLE, and in lay-by nooks, colonies of BUFF-TIPPED LUNG BUTTS. Here also we find the first GOBLINS FINGERSTALLS peering through the grasses, waxy gold or translucent pink . . .'

Our 'Alien Hunts' were of dubious scientific value, and doubtless had something in common with rarity twitching amongst birders. But more, I felt, given the hunters' fascination with plant origins and the archaeology of the tips, with the world of the metal detectorist. The outings were certainly sense sharpening, and a fascinating eye-opener about the traffic of plants. And maybe they weren't only of interest to botanical geeks. On one Saturday trip to Barking Tip we found we were sharing the territory with a gang of travellers' kids, out scouring the garbage for more substantial trophies. They latched on to us, besieging us with questions and yanking up plants for identification. After a while they fanned out independently and channelled specimens back to us through their gang leader. One of the younger children thrust him a sprig of the oriental spice plant ajowan, scientifically known as *Trachyspermum ammi*. It's a difficult umbellifer, and was quite new to me, but no longer a stranger to this burgeoning young taxonomist. 'Ammy,' he proclaimed, in his best Cockney Latin. 'Nah, they've got that already.'

Maybe it would be a misnomer to call these tip plants weeds, in the sense of plants in the wrong place. As outcasts they were precisely in the *right* place, effectively quarantined from the economic activity that had brought them to this country. Their life-spans were short. They were tipped from the lorries one month, most probably buried under new soil the next. Few even had the chance to set seed, let alone spread and cause trouble in the world from which they'd been cast out. Tips were the end of the story for the individuals that grew there, not the beachhead for some new alien invasion.

Years later, I'm still intrigued by the movements and infiltrations of weeds. Their penetration can be awesome. I have found small shoots of bittercress growing close to the electric lights inside caves, rooted in cracks in the wet limestone, and seen buddleia bushes on roofs three storeys up in Bristol. High up inside the Humid Tropics Biome at Cornwall's Eden Project I once spotted a furze bush, sprouting in the imported stone substrate, which had insinuated itself into this not-quite-hermetically sealed environment from the abundance that grow outside the biomes.

Opportunism is everything. I never saw the infamous tree-of-heaven seedling that lifted the lid of an unemptied London refuse bin during the dustman's strike in 1973. But I have glimpsed the forest of young saplings that crowd the embankments of the Circle Line, where it runs above ground. Tree-of-heaven was first grown from seed at the Chelsea Physic Garden in 1751, and the seed-keys are readily blown about by the wind, and in the slipstream of passing tube-trains. Across parts of southern Europe it is now thoroughly naturalised and seems part of the indigenous *maquis*.

The insinuation of alien species can be exquisitely intimate and precise. I often see small patches of bright blue lobelia in city pavement cracks – seeded from hanging baskets directly above. Or it can be widespread and confusing, where the expansive globetrotting of the world's most successful weeds meets botanical science's more ponderous attempts to describe it. I once found a single plant of what is currently known as Bermuda-buttercup, in brilliant lemon-yellow flower, in the tub housing a sculpture by a South American artist outside a gallery at Sintra in Portugal. 'Bermuda'-buttercup is in fact an oxalis from South Africa (where it's known as Cape cowslip) and is now a rampant weed across the globe in Scilly Isle bulbfields, Australian citrus orchards, and the Caribbean plantations where it picked up its popular name. Its bewildering nomenclature seems like a metaphor for the proliferating and protean qualities of weeds themselves.

Just occasionally the outlandish enterprise of weeds – such sharp and fast indices of change – can truly lift your heart. I visited Greenham Common not long after it was decommissioned as a nuclear base in the 1990s. The recolonisation by nature had already begun. Bats were roosting in the missile silos and natterjack toads hiding out under old ammo boxes. The grassland alongside the giant runway was already beginning to be colonised by wild flowers. What was remarkable, given that Greenham Common lies over acid greensand soils, was that many of these were chalk-loving species, led by that stout-hearted weed of railway banks and old quarries, the wild carrot, with seedheads in-folded like a hummingbird's nest. The cement in the runway was leaching into the soil, turning it alkaline. Even before it was dug up, the highway to hell was already starting to dissolve, and turning into a meadow.

The serious lesson that I've learned from three decades of stalking alien weeds in Britain is that for most of them life is sweet, but short. For the hundreds of new species that arrive each year, the available niches are small, the climate hostile, the pace of environmental change often faster than even their rapid life cycles, and most of the non-cultivated land surface already occupied by ancient and determined natives. (In this sense invasive aliens fit well into the model of weeds as plants which thrive in an ecological vacuum – in this case an absence of the predators, diseases and defensive biochemistry of their native habitats.) Only a few newcomers survive to become denizens of the wider landscape and even fewer to become nuisances. The refuse tip is a metaphor for the fate of the remainder on this larger stage. They can flare brilliantly in what is essentially a botanical freak show, but the chances of their escaping rapid annihilation and going on to become successful invaders is small. Most of the foreign species that have become naturalised have travelled, as we've seen, along a different pathway: accepted into gardens, cherished and propagated and distributed from gardener to fellow gardener, until their population reaches a level where spontaneous escape, or deliberate banishment, is inevitable.

This is the way that the dozen or so 'invasive non-native species' that worry British conservationists have entered the wild. Eight are aquatics, throw-outs from aquaria and ornamental ponds. Species such as New Zealand pigmyweed, Canadian waterweed, and parrot's feather (from South

America) form dense blankets of leaves over the surface of the water that can choke other water plants (and sometimes water animals as well). Unlike terrestrial plants they have no grounded root systems to restrict their growth.

The next most serious weed is probably rhododendron which, unusually, has the ability to invade existing ancient woodland, especially in the west of Britain. DNA analysis has shown that the commonest and most prolific variety in the wild is a complex hybrid from several garden species. It has not existed on the earth before and has few natural controls. No insects eat it, and even in cool summers that aren't good for seeding it spreads prolifically by suckers. In the unique Atlantic oakwoods along the Scottish west coast (sometimes called 'the Celtic rainforest') rhododendrons can climb the wind-stunted oaks and shade out the scarce lichens and mosses for which these woods are a last and internationally important refuge.

Indian, or Himalayan, balsam is the most popular invader with conservationists in the same way that the fox is the favourite animal of fox-hunters. Work parties spent 'balsam-bashing', despite (or perhaps because of) the pervasive scent of the squashed stems and the ceaseless grapeshot of seeds, are high points in the social calendars of conservation volunteers. Whether these jollies are justified, or have any ecological impact, are moot points.

Indian balsam was introduced to Britain from the Himalayas in 1839, as an ornamental for damp garden corners, and by the end of the century was beginning to be widely naturalised, especially along West Country rivers. In 1901 the botanist A. O. Hume found it in the Looe valley in Cornwall, and wrote one of the first and most vivid descriptions of the plant. He emphasises that, at least while it was still relatively uncommon, it was regarded as a striking and beautiful plant:

> Growing in the warm south west, with the base of the stem in the clear running stream, it is a magnificent plant, 5 to 7 feet or more in height, stalwart, with a stem 1 to 2 inches in diameter just above the surface of the water, erect, symmetrical in shape, with numerous aggregations of blossom, the central mass as big as a man's head . . . masses of bloom varying on different plants through a dozen lovely shades of colour from the very palest pink imaginable to the deepest claret, and with a profusion of large, elegant, dark green, and lanceolate leaves, some of them fully 15 inches in length.

The spread of Indian balsam dramatically accelerated in the 1950s, and by the 1980s it was growing here and there along most river systems, but still most densely in the west. The reason for its spread is the same as for all balsams: its seeds are fired off like catapult shots from the taut, elastic pods (their graphic family name is *Impatiens*), and float off on the water. Indian balsam is the tallest and fastest-growing annual weed in Britain and its biggest stands are a dramatic sight, often blanketing riversides for hundreds of yards, in rippling cloaks of purples and pinks. It was well-enough known by the late twentieth century to have picked up some popular folklore. Anne Stevenson, in her beautiful love-poem to the plant, countered the conservationists' likening of its scent to lavatory cleaner by talking of its 'ripe smell of peaches, like a girl's breath through lipstick'. Vernacular names began to appear – policeman's helmet and poor-man's orchid (from the shape of the flower), jumping jacks (from the exploding seed pods). One Somerset correspondent has told me of another tangy local name, and a story that will not endear her to conservationists in the neighbourhood of Dulverton: '[Indian balsam is] my favourite plant. I am not quite sure if it is indigenous to this area, as I was given mine from somebody's garden. I have thrown the seeds in various hedges and ditches while walking the dog, so it soon will be! . . . I was told that it was known locally as bee-bums and having noticed that bumblebees are attracted to it, and that their bums are all that is seen of them while they are on it owing to the shape of the flower, I thought it an extremely good name.'

I've seen bumblebees at work on bee-bums while I was drifting down the Exe in a boat one autumn, through stands so thick they gave the whole river an extraordinary Asiatic aura. The balsam was covered with orb-weaver spiders' webs, and the air thick with bumblebees, some of them barely able to fly because their abdomens were so loaded with yellow pollen. Whatever else Indian balsam does it is a boon to indigenous invertebrates in the season of fading warmth. But there were not many other plants visible, and my companions were insistent that the balsam was blanketing out and destroying the native riverside flora.

But a decade on, I'm not convinced that the story is as simple as a merciless slaying of the poor natives by hordes of foreign giants. Indian balsam is an annual. It may grow ten feet tall in a single summer, but it dies in winter, along with its roots. It is biologically incapable of establishing a permanently rooted population. As with most opportunist annual weeds it survives and spreads by invading open areas of bare mud and gaps between existing riverside plants, especially close to towns and cities. And in many parts of the country it's clear that the greatest encouragement to balsam's spread comes from the invasive mechanical dredging of river edges. It may be that, over a period of time, the summer shading by the balsam's foliage leads to a weakening of the native flora beneath. (Though this does not happen, for example, with the rich communities of spring-flowering perennials – orchids, ragged-robin, marsh valerian – that live in fens of saw-sedge, a species which, like Indian balsam, can reach more than six feet high in summer.) But I have yet to see a site where Indian balsam has convincingly *displaced* native plant communities. Up here in East Anglia it seems unable to even enter, let alone invade, stable reed-beds at the edges of rivers.

Indian balsam can make rural riversides ecologically monotonous and urban wastelands exotically diverse. There is no clear and easy verdict to be made about its presence in Britain. I would simply echo the warning in the final stanza of Anne Stevenson's balsam poem of the dangers inherent in the stifling – at a time of great environmental change and uncertainty – of any kind of hopeful green life:

> Love, it was you who said, 'Murder the killer
> we have to call life and we'd be a bare planet
>    under a dead sun.'
> Then I loved you with the usual soft lust of
>    October
> that says 'yes' to the coming winter and a
>    summoning odour of balsam.

Unequivocally, though, Japanese knotweed is the

invader with which a truly serious reckoning has to be made. It isn't an excuse for sociable weekend pulling parties with the Pony Club or local conservation corps. It isn't a statuesque biennial like giant hogweed, which pretty much stays where it is and dies off after flowering. Japweed, when it has settled somewhere it likes, advances by more than twenty feet a year. Its perennial roots swell and can suffocate the entire root system of other species. Japanese knotweed is not regarded as just a particularly virulent example of ordinary weedage. It's seen – and treated as – a biohazard, a vegetable pandemic. And its major cultural impact has been to create an entire industry of opposition.

Its early history is much like those of many other invasive weeds. In its native regions in Japan and northern China its natural habitats (typically for a potential weed) are disturbed areas such as river gravels and mountain scree. It's one of the first species to colonise settled volcanic lava, and can tolerate extreme levels of acidity and mineral pollution. It was brought to Europe in the mid-nineteenth century and became a great favourite in garden shrubberies, for the delicate and very oriental way its heart-shaped leaves lie in almost flat layers, and for the tassels of cream flowers which stream from them. It received the accolade of the garden designer William Robinson in 1870. In the early twentieth century a Sheffield miner was told how his father had bought a plant, and how friends were invited round to marvel at the speckled stems and graceful foliage, and how later the plant was divided up for exchange.

As usual, botanists' notes provide a map of Japweed's inexorable spread. It was first noted in the wild in London in 1900, and had reached a rubbish tip in Langley, Bucks, two years later. It was in Exeter by 1908 and Suffolk by 1924. In Cornwall in the 1930s, it earned the nickname Hancock's curse, having spread from the garden of someone with that name; and there's a story that a house in the same area was reduced in price by £100 because it was overrun with knotweed. By the 1960s its colonies had spread across Britain from Land's End to the northern tip of the Isle of Lewis.

By this time it was becoming obvious that knotweed was aggressively colonial like no other alien weed in Britain. Each plant had a root system that could extend six feet underground and advance the growing shoots of the plant twenty feet or more each year. This is the chief means by which it spreads. All the plants in Britain are female, possibly originating from just a few clones, so seed formation doesn't occur (except through hybridisation with male plants of related species, such as Russian vine). As with bindweed and ground elder, small severed sections of root can quickly begin to form new plants. Freshly cut stem fragments, placed in water in greenhouse conditions, begin to produce shoots and new roots in only six days. One more thing. The spring shoots, several of which appear from the bulky rhizome, can break through asphalt, lift concrete slabs into the air, and reach a height of five feet in four weeks.

It's no wonder that knotweed began to be regarded as trouble. From garden throw-outs it advanced along railway embankments, riversides, disturbed footpath edges and roadside ditches. It invaded churchyards and cemeteries. In places it began to form immense thickets that were as impenetrable to humans as they were smothering to the vegetation beneath. And unlike the majority of perennial weeds its roots seemed able to overwhelm those of other rhizomatous species, even bracken. But mercifully it's so far shown no propensity to invade ancient wood or grassland, or any signs of becoming an agricultural weed.

The resistance began in 1981, in the same kind of pattern as the preparation for a biohazard spill or serious outbreak of flu. Japanese knotweed was added to the Wildlife and Countryside Act of 1981, and it became 'an offence to plant or otherwise grow the species in the wild'. Under the Environmental Protection Act of 1990, Japanese knotweed 'body parts' (i.e. cuttings and excavated roots) became classified as 'controlled waste' and as such had to be disposed with at licensed landfill sites according to EPA regulations.

The availability of public money for Japanese knotweed control encouraged the growth of an entire industry dedicated to its eradication. Firms flourished lists of their customers – a mixture typically of city councils, big construction firms and estate agents – on lavish websites. Conferences were held, and manuals issued to brief local landowners and worried householders about the procedures for dealing with this demonic plant. At times they read more like the rites of exorcism than weeding instructions. There is a strict protocol. Things must be done in a proper order, at the correct phase of the year. Contamination is a real danger, and protective rituals must be followed. Simple digging up is not advisable, since this merely spreads the root fragments about. Deep excavation is the recommended option for large-scale sites, with contaminated soil being taken to a licensed disposal point. Barriers – literal fences against the invader – can be sunk in the soil as a further insurance. On smaller sites the plants can be sprayed with herbicide when they are one metre tall (usually in May) and the leaves large enough to guarantee a good uptake of chemicals. Spraying late in the season, before the knotweed shows any signs of dying off, is also regarded as effective, but 'it is necessary to protect bees and other insect visitors to the flowers'. On more sensitive sites, repeated cutting, mowing or grazing between spring and autumn helps, but all the cut material must be safely disposed of, preferably by burning on site.

The eradication programme accelerated in the spring of 2009, when a clause in the Finance Bill allowed tax relief of 150 per cent on the costs of removing of knotweed from 'contaminated' land. It's no surprise that companies took advantage of this windfall and pushed the price for clearance above £50 per square metre. By early in 2010 the annual cost of Japanese knotweed nationally had risen to more than £150 million.

But that spring a cheaper, safer and less obsessive control option drifted into view. Scientists working for CABI, a not-for-profit agricultural research organisation, announced that they had been experimenting with one of the natural predators of knotweed in its native Japan, a tiny sap-sucking insect called *Aphalara itadori*, and had obtained permission from the government to conduct field trials at unspecified sites. Mindful perhaps of the prickly-pear saga in Australia, and the possible side-effects of bio-logical control methods, the researchers will be isolating the trial plots as best as they can, and include in them wild British species that are related to Japanese knotweed (e.g. knotgrass and sorrel) to check that *Aphalara* does not target them as well.

~

Not everyone agrees that the threat from invasive immigrants – even the terrible triplets: giant hogweed, Indian balsam and Japanese knotweed – is really as serious as their opponents insist. Might it instead be a kind of alien panic based on skewed data? In 2010 David Pearman, co-editor of the seminal work on UK plant distribution, *New Atlas of the British and Irish Flora* (2009), and fellow botanist Kevin Walker looked at the occurrence of invasive plants in a new way – or rather using a different scale. They suggested, first of all, that we may get a false impression of their presence because of their 'urbanity': 'We perceive aliens to be more "common" than they actually are because many are ubiquitous close to where we live. We therefore naively assume that they are common everywhere.' We extrapolate from their abundance in disturbed urban areas – roadsides, canals, derelict industrial sites – to suggest similar densities in the wider countryside. So the authors performed what we might call a reverse extrapolation. They took figures for the abundance of invasive weeds mapped according to the normal grid unit of the 'hectad', or 10 × 10 km square, and then looked at how abundant these species were mapped at a much finer scale – in 'tetrads', or 2 × 2 km squares, inside these hectads. (The total database was huge: a total of 4 million tetrad records of 4,400 species). The disparity between the figures is striking. Japanese knotweed occurred in 83 per cent of the hectads analysed, but only 29 per cent of the tetrads in those larger units. The comparable figures for rhododendron were 70 and 22 per cent, for Indian balsam 76 and 22 per cent and giant hogweed 34 and 6 per cent. The apparent abundance of these plants seems to be concentrated into comparatively small units of the landscape, most of them in urban or suburban areas. When Pearman and Walker looked with the same level of minute attention at the countryside, and especially its ecologically richest parts, invasive aliens were, in their words, usually 'vanishingly rare'. They mapped the occurrence of the arch-villain, Japanese knotweed, in Dorset, one of Britain's most rural counties. It occurred in every single hectad, suggesting a pervasive and present danger. But in the county's 1,280 Sites of Nature Conservation Importance, it occurred in just eight.

There is no reason for complacency about aliens, as the evidence from Florida and Australia demonstrates. But like all weeds, exotic invaders only really thrive where there is disturbance, or an absence of long-established plant communities whose complex, space-demanding root systems and anciently negotiated chemical relationships can usually repel boarders. Sow the seeds of giant hogweed in the shade of an ancient wood and they will probably not even germinate, just as a modest and restrained rosebay from a Yorkshire ghyll will swell in stature and ramp away when introduced to the rich and tilled spaces of an herbaceous border. A tendency to weediness in a plant is as much a matter of opportunity as a fixed character trait.

But the opportunities are always available, somewhere, at some moment, and the alien invaders are here to stay, even if we sometimes exaggerate their threats. Occasionally they prosper because they are already too deeply entrenched: attempting to eradicate Japanese knotweed entirely would bankrupt the nation. Sometimes because the traffic in plants is in a state of continuous flux: every new crop – soya bean, statice, borage – brings new weed familiars in its wake. And sometimes simply because we love them. However much the horse-chestnut – an alien introduced from the Balkans no more than four centuries ago – is devastated by equally foreign tree diseases, it will be defended to the last ditch as the defining symbol of the English village green. Alien plants have contributed much to our national and local cultures, and sometimes to local ecologies. And they may need to make even greater contributions in the future. Climate change poses real threats to native plants. They may be driven from their traditional ranges – their comfort zones – into botanical ghettoes, even into local extinction. They will leave behind vegetational vacuums which can only be filled by plants more suited to the new conditions. Some of these may be species from the warmer south, and to rebuff them just because they are not ancestral natives is to risk increasing impoverishment of our small archipelago's flora.

So I'm wondering if there might be a less dogmatic way of evaluating invasive aliens, which takes the possibility of their positive contributions into account. I'm attracted to the concept of 'naturalisation' as a rough-and-ready index of their acceptability. In strictly scientific terms naturalisation means that an exotic species has become sufficiently well established in the wild to propagate itself and spread without deliberate human assistance. There are nearly 1,500 of these in Britain, about 150 'archaeophytes' which arrived before 1500, and another 1,300 'neophytes' which were introduced subsequently.

The cultural meanings of 'naturalisation' may also have helpful undertones, in that they emphasise acceptability and cultural 'fit'. To become naturalised is, in the dictionary definition, 'to be admitted to the rights and privileges of a native-born subject or citizen', or 'to settle down in a natural manner'. During the French Revolution, the Assembly pronounced that 'We, by act of Assembly, "naturalise" the chief Foreign Friends of Humanity'. In these definitions there is a sense of give and take, of the foreigner contributing something to its adopted culture, as well as attempting to blend in.

How does this work with some well-known alien plants? Some, like the horse chestnut (a neophyte), are already universally accepted, and thoroughly naturalised in the cultural sense. The conker tree has contributed hugely to the English landscape, to children's games, even to the cosmetic industry (the fruits produce a detergent for shampoos), while doing no more harm than unobtrusively seeding itself in a few native woodlands. Snowdrop (an archaeophyte in all but a few parts of western Britain), is another. It is highly invasive and capable of developing large colonies outside the gardens from which it escapes. But most people would be shocked to hear it described as anything other than an impeccably native English wild flower.

Large numbers of exotic garden plants have naturalised amicably and often delightfully amongst our native species. I've already mentioned the contributions made by ivy-leaved toadflax, Michaelmas daisy, green alkanet and wallflower. Some of the best newcomers originated in the Mediterranean – dame's violet, for instance, with its carnation-scented flowers; winter aconite, widely known as 'choirboys' for the green ruff of sepals that surrounds the yellow flowers; and red valerian, whose brilliant red flowers look so vivacious on stonework in south-west England.

Slender speedwell was introduced from the Caucasus as a rock-garden speciality in the 1830s, and now forms glistening pools of blue (where it is permitted) in rough lawns, amongst the buttercups and fallen cherry blossom. Tree-lupin came from California in 1793, and is naturalised on sandy shorelines along the east and south coasts, where its yellow flowers scent the sea breezes with honey in high summer. Duke of Argyll's teaplant, *Lycium barbarum*, which scrambles unobtrusively about hedges throughout southern Britain, is worth having just for the story behind it. In the early eighteenth century Archibald Campbell, third Duke of Argyll and a famous plant collector, was sent a true tea-plant, *Camellia sinensis* and a *Lycium* from China. But their labels were muddled up, and either unwittingly or as a joke he continued to grow them on under their wrong names. The story surfaced in 1838, long after the duke had died, but the plant has continued to keep its joshing title.

More contentious are the hordes of blowsy garden daffodil varieties that are now naturalised on roadsides and grassland throughout Britain. Many people welcome any splash of colour after winter, and local councils and amenity groups, doubtless with the best of intentions, are planting drifts of daffs on roundabouts and road verges and greens way outside their parish boundaries. (They can cause trouble even here. The Suffolk County Plant Recorder, Martin Sanford, has pointed out that though they're justified as 'brightening up the place', 'they are also used to stake a claim on unenclosed land and can be a precursor to more formal gardening on such sites'.) Surprisingly many of the brash, modern cultivars and hybrids seem able to set fertile seed, and colonies are cropping up in woods and field-edges miles from human habitation. I have found clumps in the middle of fens in the Norfolk Broads and by the side of a beck a thousand feet up in the Yorkshire Dales.

As daffodil foliage dies back in late May, they're so far only a minor threat to native plants (though in April they can blanket out cowslips and primroses). But in natural landscapes they are the most aesthetically pernicious of weeds, jarring with the wild native vegetation far more than vilified plants like Indian balsam, and, thoroughly out of place, carry the aura of the municipal park deep into the countryside.

At the other extreme there are invasive species which ought never to get their naturalisation papers. Parrot's feather, for example, a filigree aquarium throw-out from South America that is overwhelming pools and open areas of marshy ground. Japanese knotweed, rhododendron and cherry laurel probably. Or, much closer to home, the British-bred oil-seed rape that is increasingly invading roadside verges outside the fields where it's grown. Botanical naturalisation – the granting of honorary native status – ought to depend on acceptable and appropriate behaviour, not country of origin.

The problems arise with borderline cases, where behaviour or distribution can suddenly change. Alexanders is (or was) a much-loved and ancient archaeophyte. It was brought to Britain from the Mediterranean coast by the Romans for use as a pot-herb, and was widely cultivated in gardens until the arrival of celery, whose flavour it slightly resembles. It soon became widely naturalised (in both cultural and ecological senses) on waste places near the sea, seeming, like many Mediterranean species, to like the milder winters and saline air of the coast. Its bright viridian leaves and yellow flower tufts are amongst the first to appear on roadsides in spring, and its blanched stalks still make very passable eating. But after 2,000 years as a welcome and restrained coastal clarion, it suddenly went on the offensive, and in the mid-1990s it began spreading rapidly inland. It was thought at first that this might be a consequence of the massive increase in salt concentrations along roadside verges, as with Danish scurvy-grass, but the new colonies of alexanders are chiefly appearing along unsalted rural roads, and at the top of high banks, not close to the road surface, so some other factor – perhaps climatic – must be involved. By 2000 it was cropping up along major roads in southern Hertfordshire, more than a hundred miles from the sea. And at the same time, having colonised most of the roadsides round the north-eastern Norfolk Broads, further colonies began crowding along the A143, in the extreme south of the county. The revelation has caused a minor turf war in the county's naturalists' society, which had just voted alexanders Norfolk's County Flower. Not for the first time the image of a weed hovers between beloved symbol and incipient thug.

Images of weeds change when they move or enlarge their estate, when they make new encroachments, and shift too with alterations in popular sensibility. Just as the once abundant house sparrow, which had a bounty on its head a century ago, is now a rare and revered garden prodigy, so the image of alexanders may soon degenerate, from that of the cherished first green tufting of the year and a living piece of botanical history to a pushy immigrant taking up space that should be occupied by hard-grafting

British plants. Weeds can be plants finding themselves in the wrong moment as well as the wrong place.

Looking at another much demonised alien, the sycamore, through the lens of cultural naturalisation is revealing. The mythology stacked against it is formidable. It's a true weed, invasive and loutish. Its myriad seedlings swamp the ground and out-compete native trees. Its large and ungainly leaves litter the earth, then slowly rot to a slithery mulch. They are 'the wrong sort of leaves' that famously cause trains to skid to a halt every autumn. In September 2009 National Express East Anglia issued a special 'Autumn leaf-fall' railway timetable, headed by a photograph of a sycamore leaf.

Most of the same accusations could, of course, be levelled against many native trees – ash for its prolific seedlings, beech for its interminably slow-rotting leaf litter. But sycamore has that extra stain, the indelible mark of the vegetable beast. It is a foreigner. It was supposedly introduced from central Europe, by unknown agencies, in the sixteenth century – probably just before John Evelyn, writing from the Home Counties in 1664, gave the first version of a now familiar diatribe: 'The *Sycomor* . . . is much more in reputation for its *shade* than it deserves, for the *Hony-dew* leaves, which fall early . . . turn to a *Mucilage* and noxious *insects*, and putrifie with the first moisture of the season; so as they contaminate and marr our *Walks*; and are therefore by my consent to be banish'd from all curious *Gardens* and *Avenues*.'

A correspondent writing to me three centuries later gives a more upbeat view of its contribution to the suburban landscape:

The sycamore is often called a weed species. It certainly has one weed quality; it is a prolific seeder when mature. The winged, aerodynamic fruits are known as 'lock-and-keys'. If you would like a wood in your garden, then a nearby sycamore will oblige, without your lifting a finger . . . Rattling into London along the suburban rail lines here it is again, defying the diesel fumes on trackside embankments and urban wasteland. The adaptability of the tree in Britain is so complete it is almost comic. Here is a Johnny-come-lately outnativing the natives in almost every situation.

But is sycamore really a 'Johnny-come-lately'? There is an unmistakable carving of its leaf, displayed alongside field maple (presumed native) and greater celandine (presumed alien) on St Frideswide's shrine in Christ Church Cathedral, Oxford, which dates from 1282. The carving may, of course, have been made by a peripatetic stonemason who had seen the tree in the wild on the Continent, but it introduces a sliver of doubt about sycamore's alien origins. And in 2005 Ted Green, one of the great authorities on tree history, published a provocative paper, arguing (in the same spirit as Anne Stevenson's poem on Indian balsam) that climate change ought to make us 'rethink the relevance of taking a dogmatic position on native versus non-native [i.e. weed] trees'. He puts the case – perhaps as devil's advocate – for considering the 'Celtic maple' or 'Scots plane' a possible native. He points out that sycamore pollen may be present, unidentified, in archaeological deposits, since it is indistinguishable from field maple; that it's the equal of ash, elm and oak as a host to old-forest lichens. And, most interestingly, that a possible historical retreat – and a much more modern advance – may be due to climate change and the tree's evolving ways of dealing with this. Sooty bark disease, the greatest scourge of sycamore, has waxed and waned over the past millennia. 'Is Sycamore's ability to establish rapidly and recolonise', Green wonders, 'an evolutionary survival strategy to counter its unpredictable demise from diseases in times of heatwaves and droughts?'

The passage of time is often the deciding factor in the fortunes of invasive plants. The weeds of cultivation can only hang on to their territory if it is constantly disturbed. The rampant advances of aliens across the land continue only until some insect or microbe learns to eat them. As for sycamore stands, they have rarely been given the chance to develop to maturity, to see whether that mellows their behaviour. Where they have, they behave remarkably like native woods. I found one by chance in the Chilterns recently. It was about 150 years old, and had sprung up in a storm gap in a beechwood. It was a gracious and airy place, and the ground – far from being buried, according to popular myth, under a thick cake of half-rotted leaves – was covered with ancient woodland flowers: bluebells, wood sorrel and wood anemones. And here, at least, the weed tree wasn't regenerating under its own canopy. From the few seedlings that were evident, the next woodland tree here, in a few centuries' time, will be the indisputably native ash.

# 12

## *The Shoreditch Orchid*

And when the millennium crumbles,
I'll be squinting through the corrugated fence ...
and as I kick an old kerbstone
I'll find you, Shoreditch orchid, true and shy,
Rooting in the meadow streets
Through old cable, broken porcelain, rivets and springs;
Living off the bones of the railway.

From 'The Shoreditch Orchid' by Peter Daniels

IT'S NOT HARD to find abundant evidence even in Britain of the weed dystopias discussed in the previous chapters. To see, on any roadside verge, Stephen Meyer's glum prediction of a world of 'adaptive generalists' already becoming a reality, the cowslips and primroses replaced by goosegrass and nettle and all the rank growths of over-fertilised grassland. 'Life', Meyer wrote, 'will just be different: much less diverse, much less exotic, much more predictable.' Or to find, in the corners of any farm struggling with the recession, Neal Ascherson's abandoned agricultural machines being swallowed by brambles and hemlock. Or to walk on a Cornish cliff and see the native thrift and spring squill overwhelmed by drifts (albeit very beautiful drifts) of South African Hottentot-figs. Is this our own 'feral future', a glimpse of what weeds will continue to do to us, and to the ancient and intimate floral neighbourhoods we have lived with for millennia? Or is it what *we* are continuing to do to them?

It is even possible to enter Alan Weisman's world suddenly made free of us, to see a modern house vacated a couple of years earlier already being cracked open by Japanese knotweed and young elder bushes. It is a chastening antidote to our technological cockiness. But you can also see such places a few decades on, and the prospect of the dominion of weeds being our planet's irreversible destiny begins to waver a little. I'm in such a place as I write this, the relics of an abandoned human settlement in Essex. It was last lived in less than thirty years ago, but already its remains are less substantial than some Inca city swallowed by the forest 1,000 years before. There are a few invasive weeds – patches of goat's-rue and goldenrod, cotoneaster bushes lurking in the scrub – but mostly I feel as if I'm meandering through a wild, if slightly incoherent, juvenile woodland. There are thick stands of hawthorns and a scatter of ash trees. One of them is wreathed with thick climbers and pocked with woodpecker probings, and it takes a sudden change in the light to make me realise I'm gazing at an ancient telegraph pole, trailing lianas of rusting cable. Farther down the track there's a redundant pillar-box, another communication station closed down by the burgeoning clumps of privet and elder, and fire hydrants marooned in deep willow thickets. And everywhere amongst the native trees, especially in the more open areas, are small and far from invasive clumps of garden shrubs and orchard trees.

I must fill in some background here. This is the site of

the once thriving 'Plotland' community near Basildon, at its height a settlement of more than 8,500 dwellings. The Plotland movement originated during the agricultural depression at the end of the nineteenth century, when landowners began selling off their redundant land in small parcels or plots, and a rather superior shanty town began to sprout on the edge of Basildon. To begin with it consisted largely of caravans and old railway carriages, but increasingly the plotlanders put up self-build shacks and chalets (some 5,000 by the 1950s) and even full-dress brick bungalows. Crucially, every plot had space for a garden, and they were stocked with vegetables, shrubs, fruit trees and small patches of lawn.

Sadly the growth of this 'Arcadia for All' outgrew the provision of water and sewerage, and in the puritanical and bureaucratic atmosphere of the post-war era began to be looked on with distaste by the authorities. In 1949 a Development Corporation was set up, and the Plot-landers rehoused in Basildon New Town. The Corporation had virtually cleared all the shacks by the mid-1980s, and the intention was to return the area to farmland. But the ground was too steep and the soil too poor, and nobody wanted to buy. So in 1989 the Corporation took the unprecedented step of declaring the site of this ghost town a nature reserve.

What happened since has been one of the strangest and most heartening examples of natural succession. In the first few years the lawn grasses and all their accompanying weeds ran riot. The garden perennials grew to prodigious size. But then the trees started to close in, the plums and apples and taller garden shrubs. The native species – oak, ash, hawthorn and even a few of Essex's 'county tree', the hornbeam – followed, and began to sprout from bird-sown seed in the abandoned vegetable beds and lawns. Now lilac and laburnum flower amidst the may blossom. Gorgeous double roses entwine with their wild cousins, and feral grapevines scramble over the ruins of chimney breasts. Here and there in the midst of dense blackthorn scrub there is a surviving philadelphus or garden privet. But inexorably the native hardwood trees are shading out the smaller and more delicate cultivars. In the grassier areas, most of the naturalised garden plants have been replaced by native grassland species such as wild carrot and bird's-foot trefoil. In time, in the British climate, native plant communities will almost always return and supplant pioneer weeds and alien invaders.

And this is after just thirty years of growth. In another hundred the Plotlands will resemble an ordinary mixed woodland, except that it will be studded with a few sturdy apple trees and shade-tolerant box thickets and pink path-side patches of soapwort – all salutary reminders that this was once a lively human community. And a reminder too that, in temperate Britain at least, the occupation of disrupted land by weeds is rarely permanent or inexorable.

Times are changing for weeds. They are at one and the same time more successful and more brutally attacked than at any time in history. It was never the purpose of this book to examine the technical issues involved in weed control, either by farmers, gardeners or conservationists, but to look at the immensely varied motives for this control, and the impact it has had on our relationships with the plant world and with nature in general. But different technologies have different cultural side-effects. The introduction of organo-phosphorus weedkillers in the 1940s has had a greater impact on weed populations (and our attitudes towards them) than any previous technology. Chemical herbicides have been so successful that many weeds are now scarce and unfamiliar, 'passing out of our knowledge', as John Clare might have said. But in essence chemical weed killing is no different from the most primitive prehistoric hand-pulling. It's simply a way of selectively removing unwanted plants from amongst the wanted. And just as the hoe gave advantage to individual weed plants with deep roots that could regenerate when chopped, and grain sieves helped the evolution of weeds whose seeds were the same size as the crop grains, so chemical weedkillers actively encourage the evolution of individual weeds whose biochemistry is unusual enough to be immune to the poison. There was a degree of inevitability about the identity of the first weed to show resistance to organic herbicides. In Washington DC in 1990, a population of that ancient gardeners' familiar, groundsel, proved to be resistant to the widely used weedkiller simazine. Groundsel can go from seed to seeding in a period of six weeks, and produce up to five generations a year – a prodigious real-world experiment in the production of new strains. Twenty years on, fifty more species have evolved some type of resistance to a wide range of herbicides.

The cultural effect of this endless game of chemical poker, as increasingly toxic and carbon-greedy herbicides encourage the evolution of ever-more resilient weeds, has been to generate a public backlash. Many gardeners now tolerate and even enjoy the more colourful weeds in their gardens. Whole uncultivated corners may be devoted to them and the insects and birds they attract (and therefore, by putting weeds in a 'right' place, nullifying their status as weeds). Nettles, dandelions, chickweed are harvested as wild foods by the more adventurous.

Out in the countryside some farmers are developing similarly benign attitudes, leaving six-foot strips of weeds along the edges and headlands of their arable fields. This, to be fair, is chiefly for the benefit of another crop, the pheasant. But it also provides refuges for finches and barn owls and insects which help with pollination and pest control. Twenty arable weeds are now included in the UK Biodiversity Plan, which means there is government support available for increasing their populations, or establishing sanctuaries for them. Ironically the list includes species such as cornflower, pheasant's-eye and corn buttercup, which were regarded as pestilential nuisances 300 years ago.

'Integrated weed management' systems are reintroducing practices which echo those used by William Ellis in the 1700s, for instance under-sowing maize crops with a useful 'weed' like clover to suppress more harmful ones, and naturally boost available nitrogen. There are even experiments underway involving the deliberate sowing of wild weed seed in crop fields, in the hope that it will interbreed with herbicide-resistant strains of the same weed species and produce offspring more susceptible to weedkillers.

~

It is in urban areas, however, that the cultural attitude towards weeds has changed most dramatically. On derelict industrial sites and railway sidings and the margins of run-down housing estates there are no acres of productive foodland or pristine ecosystems to be ruined by the alien invaders. Such locations are dismissed as wasteland, or, in that deliberately derogatory phrase of politicians and developers, as 'brownfield sites' – an absurd misnomer which ignores the fact that these are the most jazzily colourful and biologically rich zones of the cities.

They are also places which, in the absence of competing native plant species, are seeing the evolution of authentic urban ecosystems, plant communities new to the planet but entirely appropriate for cosmopolitan built-up areas. In 1993 Oliver Gilbert, from the University of Sheffield, produced a report on these communities for the government body English Nature, called *The Flowering of the Cities*. Its subtitle – 'The natural flora of "urban commons"' – made official a phrase that had already entered the patois of urban florophiles. These backlots are unofficial commons not just because of their rowdy vegetation, but because they are the last informal open spaces left in most cities: weekend strollers pick blackberries, children build camps, travellers' settlements materialise, complete with horses and goats.

Gilbert surveyed the unique plant communities of the urban commons in several British cities, showing just how distinct their weed flora is and how related to each city's social history. Birmingham has an abundance of golden rod, probably because it is a common allotment plant, and the city has a long tradition of 'guinea gardens' and allotments. Manchester and Swansea, with their wet and

westerly climates, both have huge populations of Japanese
knotweed. In Glasgow, 'the vast site formerly occupied by
the Dalmarnock Generating Station is now being colonised [partly from moorland nearby] by an assemblage
of woody plants that include silver birch, Scots pine, grey
alder, common alder, broom, goat willow, and cotoneaster.
Nothing similar is known in other UK cities.' Sheffield,
Gilbert's own city, is one of the most colourful in Britain.
The banks of the River Don carry woodland flowers under
a spontaneously coppicing layer of Japanese knotweed
and, along with hilltops around the city, sheets of goat's-
rue, feverfew, tansy and Michaelmas daisy. Gilbert learned
from a local miner that at the beginning of the twentieth
century, hawkers used to work the poor quarters of the city
selling these barely domesticated plants as garden novel-
ties. And then there is Bristol, the *axis mundi* of Britain's
buddleia, our most popular invasive alien. Buddleia was
brought to Europe from the mountains of the Tibetan-
Chinese border in the 1880s. It is a scree plant and, like
Oxford ragwort, found the stony ballast along the railways a familiar and congenial habitat. Its light winged
seeds are drawn along in the slipstream of trains, and from
its strongholds on the railway embankments it has colonised bomb sites, walls, allotments, car parks. Sometimes
its seeds are blown upwards and take root in chimneys.
In Bristol it builds up thickets on bridges and the ledges
of buildings, sometimes forming a unique kind of urban
woodland with scrub willow and silver birch. 'Buddleia
Rules OK' (or is it UK?) reads a scrawled graffiti on one
Bristol wall.

What's clear is that for those who live in cities, the ecological and cultural profiles of these weed communities are

intricately entwined. They are recognisable neighbours, vegetable squatters, a kind of living graffiti – impertinent, streetwise, living one step ahead of the developers and the puritan fusspots. The spirit of Banksy is alive in them. In the late 1980s a run-down allotment at the back-end of Sandown on the Isle of Wight became the stage for a piece of ecological agitprop. A colony of an exceptionally rare, protected and highly local arable weed, Martin's ramping fumitory, was discovered there, and the allotmenteers used its legal status to successfully resist the development of the site. At the Barbican Gallery in the City of London, in a 2009 show entitled *Radical Nature: Art and Architecture for a Changing Planet*, the conceptual artist Simon Starling exhibited a three-dimensional cartoon, mounted on pontoons. Called *Island for Weeds*, it was a scale model of a 'quarantine raft' for invasive rhododendrons, designed to keep them alive in the wild but out of mischief, by floating them off into the middle of a loch. It reminded me of the medieval animal-defence lawyer Bartholomew Chassenée's solution for an infestation of weevils, that there should be 'set aside for the said *bestioles* alternative pasture'.

For the Norfolk and Norwich Festival of 2009, the London-based Mauritian artist Jacques Nimki produced a set of postcard sketches of city weeds, 'The Norwich Florilegium', with an off-the-wall commentary on the back of each. There were notes on names, origins, enemies, locations, favourite planting music, best mates. Imagine a slightly inebriated John Clare, in sharpest botanical mood, doing stand-up comedy for a knowing Norwich audience. All the plants busked by Nimki on his cards are real (but have invented names), are beautifully sketched, and – again recalling Clare – are exactly placed inside the intimate geography of the city.

Here is 'Winy Jack', more usually known as hedge mustard. 'Location: 52 can be found against the blue hoarding in King Street – this is where they hang out as a gang. The best examples in Norwich. Enemies: Photosystem II inhibitor. Best mates: In Norwich sometimes seen with Red Cheesebowls or Stinker Bob, and maybe Swutchen, but can be a bit territorial.' Or 'Dindle', aka nipplewort. 'Location: They can be found all over the city. The best example can be seen at 32 London Street, just over the shop, on the left-hand side. If you look at the Google satellite map, just past Swan Lane you can see the shadow of the plant on the pavement. The shop is due for a paint job, so if the plant's gone, have a look next door at 34, again on top of the shop. Two further beauties can be seen on the wall behind the bike shelter at Norwich rail station. Uses: Absolutely superb invasive plant. Tough to get rid of, known as one of the all-time greats.'

The same year, a hundred miles south in Deptford, east London, a group called the Pink Posse (the name is a botanical in-joke: the so-called 'Deptford pink' has never grown in the borough and was a misidentification by John Gerard) took the idea of weeds as graffiti even further. They began 'tagging' lively clumps of weeds on the pavements and old walls of Deptford – stencilling their names on nearby kerbstones or old bricks. They saw the weeds as tagged to their habitats, too, and have published a Google map of their 'tagworts' with links to notes about the plants. They don't have the surreal wit of Nimki's 'Florilegium', but some nice touches that will make the inquisitive look more closely at these plants done up with the weed's equivalent of a memorial Blue Plaque: 'Pellitory-of-the-wall. A wall lover, this species tags older walls all over London. It will also grow on stony ground and on gorgeous brownfield sites. Old French and Latin are the origins of the words Pellitory and Parietaria [the scientific name] – they mean wall. So perhaps we should call it Wall-of-the Wall or just Walltagger.'

~

A triangle formed by joining Deptford with the haunts of Shoreditch's imaginary orchid and the Walthamstow marshes that were the north-eastern limit of my 1970s urban ramblings would contain a host of charged botanical nodes. The relics of the bombed churches that Rose Macaulay wandered round like a mesmerised dowser. The Tower, one of the last remaining refuges for that phoenix of the Great Fire of 1666, London rocket. The shiny new marinas round Wapping and Limehouse, their walls still draped with the dye-plant gipsywort which an earlier generation of City wide-boys used to black up their faces, hoping to give the impression they were predictors of the future from the East. A few lingering rubbish tips, barely given the chance to vegetate now, let alone grow a full-blown hash plant. And out near Stratford Marsh, the canal bank where I had the most thrilling urban trek of my life. Along the Hackney Cut the towpath had turned into a jungle track, half-blocked by festoons of Russian vine. The banks were covered with drifts of dwarf elder, aka Danewort, leaves smelling of beef-gravy and covered with rosettes of pink-tipped flowers, but already tinged blood-red in the stems. (In folklore this plant is supposed to grow from the blood of Danes defeated in battle with the English.) There were heraldic towers of giant hogweed and scattered tumps of one of London's weed specialities: bladder-senna from the Mediterranean, with its huge-inflated seed pods, which blue tits were ripping open and squeezing their heads into, in search of grubs.

Now the whole of this patch, much of Stratford Marsh and the Lower Lea valley, the epicentre of east London's feral vegetation, has been swallowed up by the Olympic Park. Five hundred acres of rough pasture, wild scrubland, allotments, plus 2,000 houses, have been bulldozed. It is the most colossal upheaval seen in east London since the Blitz and, geologically, probably the most long-drawn-out since the Ice Age. I would love to have been able to see how weeds were coping with this man-made earthquake, the very kind of tumult they were evolved to cope with. But at the time of writing visitors are banned. Health and safety and security are the Olympic Development Authority's excuses, but I can't imagine they have a tolerant view of nosey writers on the lookout for precisely the vegetable detritus they are spending large sums of money to eradicate.

So I gaze at the Park from some way off. About once a month I take the train from Norfolk to Liverpool Street, and it passes within a few hundred yards of the nascent stadium, which seems to be rising out of the earth much as the Mother Ship in *Close Encounters of the Third Kind* settled gently down upon it. I see glints and shards of weeds, rivulets of yellow along a damp lorry track, a conical spoil tip topped with a fuzzy green skull-cap. I have no idea what they are and a month later they're gone, replaced by a new cast. What looks like Oxford ragwort is tangling with the

fences, and drifts of Danewort are edging a stretch of the
River Lea. Scandinavian contractors should beware. It's
as if I'm witnessing plant succession here in fast-motion
– except that this is weeds' *habitual* pace of life. They've
evolved for the sprint, for the landslide, for the volcanic
simulacrum of the controlled explosion. Here, though,
the speed of change is so fast that many of them do not
even get the chance to reproduce themselves. So what is
the point of their brief invasion, of the very existence of
such ephemeral organisms? When I look at their comings
and goings, as hectic as the movements of the bulldozers, I grope for metaphors to understand their meaning. I
think of ants, but they're too organised, too determinedly
earth-changing, like the excavating machinery itself. Then
it occurs to me that they are like a kind of immune system,
organisms which move in to repair damaged tissue, in this
case earth stripped of its previous vegetation.

But this doesn't mean that weeds have a 'purpose', any
more than does any other living thing. An organism exists
for no other reason than that it is able to, and can find
an opportunity to do so. The wonderful, almost transcendental thing about life on earth is that in order to so exist,
organisms must relate to each other and to the earth itself,
and therefore find, if not a purpose, something close to a
role. Weeds' rapid, opportunist lifestyles mean that their
role – what they do – is to fill the empty spaces of the earth,
to repair the vegetation shattered naturally for millions
of years by landslide and flood and forest fire, and today
degraded by aggressive farming and gross pollution. In so
doing they stabilise the soil, conserve water loss, provide
shelter for other plants and begin the process of succession to more complex and stable plant systems. I think

it's a reasonable assumption that if weed elimination had been an option when farming was first developed 10,000 years ago, agriculture would have died a quick death as a passing fancy. Once cultivated, the dry soils of the Middle East would have simply blown away. The crops would have foundered for lack of protection from the sun. Perhaps properly understood, compromised with rather than exterminated, weeds might help us now, as they seem to be doing in the experiments in ecologically friendly crop-management systems.

But reaching a rapprochement with weeds – and their inevitability – will always be a dizzying process. It involves marrying practical control with cultural acceptance. For most of recent history we have been travelling in the opposite direction, and the more we seem able effortlessly – but temporarily – to eradicate weeds the less we bother to understand them. From the beginnings of farming to the start of the Agricultural Revolution, weeds were accepted as a troubling but necessary part of natural life and human existence. They might need to be cleaned from the soil, but they were also indications of its fertility. They were employed in rituals to speed growth. They were regarded as powerful medicines, and more pragmatically as important contributors to domestic economy. The biblical insistence that they were a punishment for mankind's original sin (remembered well into the nineteenth century) carried with it a subliminal understanding that they were also an ecological punishment. They were the tithe we paid for breaking the earth.

Yet once it became possible to attack them with machines and then chemicals, weeds slipped out of our understanding. Their appearance now sparks reflexes, not reasoning.

They are regarded as inexplicable and impertinent intruders, quite unconnected with the way we live our lives. And in a radical shift of perspective we now blame the weeds, rather than ourselves. Yet we gave them their derogatory name, and the opportunity to extend their repairing role out of the wilderness and into our damaged world. Every single weed nuisance – from the ground-elder in the over-hoed English herbaceous border, to the casually imported pond plant suffocating the Everglades swamps, to the cogon smothering the napalmed remnants of the Vietnamese rainforest – has been the consequence of thoughtless and sometimes deliberate disruption of natural systems. Weeds are our most successful cultivated crop.

The American writer Carl Safina has pointed out how the albatross, the victim of a legendary slaying, has, in popular lore, mutated into a *cause* of bad luck. To have 'an albatross round your neck' is to be burdened with social baggage. It is the bloody bird's fault. In the poem that started the legend, *The Rime of the Ancient Mariner*, Coleridge suggests something quite different. The mariner dementedly shoots the albatross, 'a pious bird of good omen' . As a result his ship is becalmed, and his suffering crew hang the dead bird round his neck. To have an albatross round the neck is to be in a predicament that one has caused oneself. We get the weeds we deserve, our vegetable albatrosses.

At the start of this book I suggested that weeds were a consequence of our rigid separation of the natural world into the wild and the domestic. They are the boundary breakers, the stateless minority, who remind us that life is not that tidy. They could help us learn to live across nature's borderlines again.

# Glossary of plant names

~

Scientific names for British plants are taken from: Clive Stace, *New Flora of the British Isles*, Cambridge, 1991. For other regions: D. J. Mabberley, *The Plant Book*, 2nd edn, Cambridge, 1997. Alternative and widely used popular names are given in brackets. More local vernacular names are not given below, and are distinguished in the main text by their initial capitalisation.

adder's-tongue, *Ophioglossum vulgatum*
ajowan, *Trachyspermum ammi*
alder (common), *Alnus glutinosa*
   grey, *A. incana*
alexanders, *Smyrnium olustratum*
alfalfa (lucerne), *Medicago sativa sativa*
ash, *Fraxinus excelsior*
asparagus, *Asparagus officinalis* ssp. *officinalis*
autumn lady's-tresses, *Spiranthes spiralis*
balsam, Indian (Himalayan), *Impatiens glandulifera*
   orange, *I. capensis*
   small, *I. parviflora*
bee orchid, *Ophrys apifera*
beech, *Fagus sylvatica*
Bermuda-buttercup, *Oxalis pes-caprae*
betony, *Betonica officinalis*
bindweed, black, *Fallopia convolvulus*
   field, *Convolvulus arvensis*

hedge, *Calystegia sepium*
bird's-foot-trefoil, common, *Lotus corniculatus*
bittercress, hairy, *Cardamine hirsuta*
black mustard, *Brassica nigra*
blackcurrant, *Ribes nigrum*
blackthorn, *Prunus spinosa*
bladder-senna, *Colutea arborescens*
bluebell (English), *Hyacinthoides non-scripta*
    Spanish, *H. hispanica*
borage, *Borago officinalis*
box, *Buxus sempervirens*
bracken, *Pteridium aquilinum*
bramble (blackberry), *Rubus fruticosus*
Brazilian pepper, *Schinus terebinthifolius*
bristle-grass, *Setaria* spp.
broom, *Cytisus scoparius*
bryony, black, *Tamus communis*
    white, *Bryonia dioica*
buckwheat, *Fagopyrum esculentum*
buddleia (butterfly-bush), *Buddleja davidii*
buffel grass, *Cenchus ciliarus*
bugloss, small-flowered, *Echium parviflorum*
    viper's, *Echium vulgare*
burdock, greater, *Arctium lappa*
    lesser, *A. minus*
burnet-saxifrage, *Pimpinella saxifraga*
buttercup, *Ranunculus* spp.
    corn, *R. arvensis*
    creeping, *R. repens*
butterwort (common), *Pinguicula vulgaris*
caltrops, *Tribulus terrestris*
Canadian waterweed, *Elodea canadensis*
canary-grass, *Phalaris canariensis*
cannabis; hemp, *Cannabis sativa*
capeweed, *Arctotheca calendula*
castor-oil plant, *Ricinus communis*

celandine, lesser, *Ranunculus ficaria*
    greater, *Chelidonium majus*
chamomile (mayweed), corn, *Anthemis arvensis*
    stinking, *A. cotula*
charlock, *Sinapis arvensis*
cherry plum, *Prunus cerasifera*
cherry laurel, *Prunus laurocerasus*
chervil, garden, *Anthriscus cerefolium*
chickweed, common, *Stellaria media*
chicory, *Chicoria intybus*
cinchona, *Cinchona officinalis*
cinquefoil, creeping, *Potentilla reptans*
clover, red, *Trifolium pratense*
    sulphur, *T. ochroleucon*
    white, *T. repens*
cockspur grass, *Echinochlia crus-galli*
coco grass, *Cyperus rotundus*
cogon, *Imperata cylindica*
coltsfoot, *Tussilago farfara*
comfrey, common, *Symphytum officinale*
    white, *S. orientale*
corncockle, *Agrostemma githago*
cornflower, *Centaurea cyanus*
couch, *Eyltrigia repens*
cow parsley, *Anthriscus sylvestris*
cowslip, *Primula veris*
crab apple, *Malus sylvestris*
cuckooflower, *Cardamine pratensis*
cumin, *Cuminum cyminum*
cyclamen (sowbread), *Cyclamen hederifolium*
daisy (common), *Bellis perennis*
    oxeye, *Chrysanthemum vulgare*
dame's-violet, *Hesperis matronalis*
dandelion, *Taraxacum officinale*
darnel, *Lolium temulentum*
Deptford pink, *Dianthus armeria*
dill, *Anethum graveolens*

dock, broad-leaved, *Rumex obtusifolius*
    curled, *R. crispus*
    patience, *R. patientia*
    wood, *R. sanguineus*
dodder, *Cuscuta epithymum*
    flax, *C. epilinum*
dog-rose, *Rosa canina*
duckweed, *Lemna* spp.
Duke of Argyll's teaplant, *Lycium barbarum*
elder, *Sambucus nigra*
    dwarf, *S. ebulus*
elecampane, *Inula helenium*
emmer, *Triticum turgidum*
evening primrose, common, *Oenothera biennis*
everlasting-pea, broad-leaved, *Lathyrus latifolius*
fat-hen, *Chenopodium album*
fennel, *Foeniculum vulgare*
fenugreek, *Trigonella foenum-graecum*
feverfew, *Tanacetum parthenium*
field cow-wheat, *Melampyrum arvense*
fig, *Ficus carica*
fleabane, Canadian, *Conyza canadensis*
    Sumatran, *C. sumatrenis*
fluellen, *Kickxia* spp.
forget-me-not, *Myosotis* spp.
foxtail, meadow, *Alopecurus pratensis*
fuchsia, *Fuchsia magellanica*
fuller's teasel, *Dipsacus sativus*
fumitory, common, *Fumaria officinalis*
    Martins' ramping, *F. reuteri*
furze (gorse), *Ulex europeaus*
gallant-soldier, *Galinsoga parviflorus*
gipsywort, *Lycopus europaeus*
gladiolus, wild, *Gladiolus italicus*
goat's-rue, *Galega officinalis*
goldenrod, Canadian, *Solidago canadensis*
goosegrass (cleavers), *Galium aparine*

grama, *Bouteloua* spp.

greater stitchwort, *Stellaria holostea*

green alkanet, *Pentaglottis sempervirens*

ground-elder, *Aegopodium podagraria*

ground ivy, *Glechoma hederacea*

groundsel, *Senecio vulgaris*

hawksbeard, smooth, *Crepis capillaris*

hawkweeds, *Hieracium* spp.

hawthorn (may), *Crataegus monogyna*

heather (ling), *Calluna vulgaris*

hedge mustard, *Sisymbrium officinale*

hedge woundwort, *Stachys sylvatica*

hemlock, *Conium maculatum*

henbane, *Hyoscamus niger*

hoary cress, *Lepidium draba*

hogweed (common), *Heracleum sphondylium*
  giant, *H. mantegazzianum*

honeysuckle (woodbine), *Lonicera periclymenum*

hornbeam, *Carpinus betulus*

horse-chestnut, *Aesculus hippocastanum*

horsetail, field, *Equisetum arvense*

Hottentot-fig, *Carpobrotus edulis*

hound's-tongue, *Cynoglossum officinale*

houseleek, *Sempervivum tectorum*

ivy, *Hedera helix*

ivy-leaved toadflax, *Cymbalaria muralis*

knotgrass (ironweed), *Polygonum aviculare*

kudzu vine, *Pueraria montana*

laburnum, *Laburnum anagyroides*

lady's-bedstraw, *Galium verum*

lady's-mantle, *Alchemilla* spp.

lamb's ears, *Stachys byzantina*

lilac, *Syringa vulgaris*

linseed (flax), *Linum usitatissimum*

lobelia, *Lobelia erinus*

loganberry, *Rubus loganobaccus*

London rocket, *Sisymbrium irio*

love-lies-bleeding, *Amaranthus caudatus*
mallow, common, *Malva sylvestris*
    musk, *M. moschata*
mandrake, *Mandragora officinarum*
mare's-tail, *Hippuris vulgaris*
marigold, common, *Calendula officinalis*
    corn, *Chrysanthemum segetum*
meadow-grass, smooth, *Poa pratensis*
Michaelmas daisy, common, *Aster* x *salignus*
millet, *Panicum miliaecum*
mint, *Mentha* spp.
montbretia, *Crocosmia* x *crocosmiiflora*
morning glory, *Ipomoea purpurea*
motherwort, *Leonurus cardiaca*
mugwort, *Artemisia vulgaris*
mullein, great, *Verbascum thapsus*
navelwort (pennywort), *Umbilicus rupestris*
nettle, small, *Urtica urens*
    stinging, *U. dioica*
New Zealand pigmyweed, *Crassula helmsii*
nightshade, black, *Solanum nigrum*
    deadly, *Atropa belladonna*
    woody (bittersweet), *Solanum dulcamara*
nipplewort, *Lapsana communis*
oil-seed rape, *Brassica napus* ssp. *oleifera*
oregano (marjoram), *Origanum vulgare*
oxlip, *Primula elatior*
oxtongue, bristly, *Picris echioides*
pampas grass, *Cortaderia selloana*
pansy, field, *Viola arvensis*
    wild, *V. tricolor*
paperbark, *Melaleuca* spp.
parrot's-feather, *Myriophyllum aquaticum*
pellitory-of-the-wall, *Parietaria judaica*
penny-cress, field, *Thlaspi arvense*
peony, *Paeonia mascula*
pheasant's-eye, *Adonis annua*

pimpernel, blue, *Anagallis arvensis.* ssp. *caerulea*
    scarlet, *A. arvensis*
pineapple-weed, *Matricaria discoidea*
pirri-pirri-bur, *Acaena novae-zelandiae*
plantain, greater, *Plantago major*
    hoary, *P. media*
    ribwort, *P. lanceolata*
poison ivy, *Rhus radicans*
poppy, corn, or common, *Papaver rhoeas*
    opium, *P. somniferum*
prickly pear, *Opuntia ficus-indica*
primrose, *Primula vulgaris*
privet, garden, *Ligustrum ovalifolium*
    wild, *L. vulgare*
purple-loosestrife, *Lythrum salicaria*
quackgrass, *Agopyron repens*
quaking-grass, *Briza media*
radish, wild, *Raphanus raphanistrum*
ragged-robin, *Lychnis flos-cuculi*
ragweed, *Artemisia artemisiifolia*
ragwort, common, *Senecio jacobaea*
    London, *S.* x *subnebrodensis*
    Oxford, *S. squalidus*
red-hot poker, *Kniphofia* spp.
redshank, *Persicaria maculosa*
reed, *Phragmites australis*
rhododendron, *Rhododendron ponticum*
rosebay willowherb, *Chamerion angustifolium*
rue-leaved saxifrage, *Saxifraga tridactylites*
Russian vine, *Fallopia baldschuanica*
rye-grass, Italian, *Lolium multiflorum*
    perennial, *L. perenne*
sandwort, spring, *Minuartia verna*
saw-sedge (great fen-sedge), *Cladium mariscus*
Scots pine, *Pinus sylvestris*
scurvy-grass, common, *Cochlearia officinalis*
    Danish, *C. danica*

self-heal, *Prunella vulgaris*

shepherd's-needle, *Scanidx pecten-veneris*

shepherd's-purse, *Capsella bursa-pastoris*

shoo-fly, *Nicandra physalodes*

silver birch, *Betula pendula*

silverweed, *Potentilla anserina*

soapwort, *Saponaria officinalis*

sow-thistle, *Sonchus* spp.

soya bean, *Glycine max*

speedwell, germander, *Veronica chamaedrys*
    slender, *V. filiformis*

spiny chicory, *Cichorium spinosum*

spiny restharrow, *Ononis spinosa*

spotted medick, *Medicago arabica*

squill, spring, *Scilla verna*

St John's-wort, perforate, *Hypericum perforatum*

statice, *Limonium* spp.

stinking hellebore, *Helleborus foetidus*

striga (witchweed), usually *Striga lutea*

sun spurge, *Euphorbia helioscopia*

sunflower, *Helianthus annuus*

sycamore, *Acer pseodoplatanus*

tansy, *Tanacetum vulgare*

thistle, creeping, *Cirsium arvense*
    spear ('Scotch'), *C. vulgare*

thorn-apple, *Datura stramonium*

thoroughwort, *Ageratina adenophora*

thrift (sea-pink), *Armeria maritime*

thyme, wild, *Thymus polytrichus*

tree lupin, *Lupinus arboreus*

tree-of-heaven, *Ailanthus altissima*

tumbleweed (Russian thistle, spineless saltwort), *Salsola kali*, ssp.
    *ruthenica*

valerian, marsh, *Valeriana dioica*
    red, *Centranthus ruber*

vervain, *Verbena officinalis*

violet, sweet, *Viola odorata*

wall barley, *Hordeum murinum*
wallflower, *Erysimum cheiri*
walnut, *Juglans regia*
water hyacinth, *Eichornia crassipes*
watercress, *Rorripa nasturtium-aquaticum*
water-pepper, *Persicaria hydropiper*
weld, *Reseda luteola*
white horehound, *Marrubium vulgare*
wild carrot, *Daucus carota*
wild leek, *Allium ampeloprasum*
wild oats, *Avena fatua*
willow, goat, *Salix caprea*
    grey (sallow), *S. cinerea*
winter aconite, *Eranthis hyemalis*
winter heliotrope, *Petasites fragrans*
woad, *Isatis tinctoria*
wormwood, *Artemisia absinthium*
yarrow, *Achillea millefolium*
yellow-rattle, *Rhinanthus minor*
Yorkshire-fog, *Holcus lanatus*

# Notes and References

❧

I have not included sources below where a full reference is given in the text.

Page 1. The full story of these Middlesex explorations is in Richard Mabey, *The Unofficial Countryside*, London, 1973.

Page 7. Scottish thistle: Tim Low, *Feral Future: The Untold Story of Australia's Exotic Invaders*, Chicago, 2002. Ralph Waldo Emerson, *Fortune of the Republic*, 1878.

Page 8. Francis Simpson, *Simpson's Flora of Suffolk*, Ipswich, 1982.

Page 9. John Ruskin, *Proserpina*, 1874-6. J. C. Loudon, *Arboretum and Fruticetum Britannicum*, 1838.

Page 10. American attitudes: Paul Robbins, *Lawn People: How Grasses, Weeds and Chemicals Make Us Who We Are*, Philadelphia, 2007; Sara Stein, *My Weeds: A Gardener's Botany*, New York, 1988.

Page 13. Indian balsam games: Richard Mabey, *Flora Britannica*, London, 1996. Plants in Harry Potter: see Wikipedia.

Page 16. Loosestrife nomenclature: Geoffrey Grigson, *A Dictionary of English Plant Names*, London, 1974.

Page 18. Stephen Meyer, *The End of the Wild*, Cambridge, Mass., 2006.

Page 19. Gary Snyder, 'Earth turns', *Orion*, April 2009.

Page 20. Gerard Manley Hopkins, 'Inversnaid'. John Clare, 'Leisure', in *John Clare by Himself*, ed. Eric Robinson and David Powell, Ashington, 1996.

Page 23. Bomb-site weeds: Edward Salisbury, *Weeds and Aliens*, 2nd edn, London, 1961; Richard Fitter, *London's Natural History*, London, 1945.

Page 25. Michael Pollan, *Second Nature*, London, 1996.

Page 27. 'Flowers of the field': Lytton John Musselman, *Figs, Dates, Laurel and Myrrh: Plants of the Bible and the Quran*, Portland, 2007. John Gerard, *The Herball, or Generall Historie of Plantes*, 1597, and Dover facsimile of 1633 edn, New York, 1975. Salisbury, *Weeds and Aliens*, op. cit.

Page 29. Charles Darwin, *The Origin of Species*, 1859.

Page 31. Gilbert White's garden: Gilbert White, *Journals*, ed. Francesca Greenoak, 3 vols, London, 1987-9.

Page 35. Northumberland folk-lore, quoted in Roy Vickery, *Plant Lore*, Oxford, 1995. John Ray, *Catalogus Plantarum circa Cantabrigiam nascentium*, 1660.

Page 36. Shirley poppies: Mabey, *Flora Britannica*, op. cit.

Page 40. Genesis, see versions in Robert Gould and Stephen Prickett (eds), *The Bible: Authorised King James Version*, Oxford, 1997.

Page 41. Garden of Eden: John Prest, *The Garden of Eden: The Botanic Garden and the Re-creation of Paradise*, New Haven, 1981; Max Oelschlaeger, *The Idea of Wilderness*, New Haven, 1991.

Page 42. N. I. Vavilov, *Studies on the Origin of Cultivated Plants*, Leningrad, 1926; see also Geoffrey Grigson, 'Ninhursaga', in *Gardenage*, London, 1952. Desert food: 'Wild plants in the cuisine of modern Assyrians in the eastern Syrian-Turkish borderland', Oxford Food Symposium, 2004; Terence McKenna, *Food of the Gods*, New York, 1984.

Page 45. Purple rice: Stein, *My Weeds*, op. cit.

Page 47. John Passmore, *Man's Responsibility for Nature: Ecological Problems and Western Tradition*, New York, 1974. Virgil, *The Georgics*, ed. and trans. K. R. Mackenzie, London, 1969.

Page 48. Claude Lévi-Strauss, *From Honey to Ashes*, 1966, trans. John and Doreen Weightman, New York, 1973.

Page 49. Neolithic weeds: Phil Watson and Miles King, *Arable Plants*, Old Basing, 2003.

Page 50. Eilert Ekwall, *The Concise Oxford Dictionary of Place Names*, Oxford, 1960. P. V. Glob, *The Bog People: Iron-Age Man Preserved*, 1965 (in Danish; UK edn, London 1969).

Page 57. Durer: Madeleine Pinault, *The Painter as Naturalist*, Paris, 1991; see also Wilfrid Blunt, *The Art of Botanical Illustration*, London, 1950.

Page 58. Johann von Goethe, *The Sorrows of Young Werther*, 1774.

Page 60. Dioscorides quoted in Wilfrid Blunt and Sandra Raphael, *The Illustrated Herbal*, London, 1978.

Page 64. Foster Barham Zincke, *Some Materials for the History of Wherstead*, 2nd edn, 1893. Thomas Tusser, *His Good Points of Husbandry* (c. 1557), ed. Dorothy Hartley, London, 1931.

Page 65. Dorothy Hartley, *The Land of England*, London, 1979.

Page 68. Bindweed: Salisbury, *Weeds and Aliens*, op. cit.

Page 70. Trials of animals: Noel Sweeney, 'Animals in the dock', *BBC Wildlife*, March 2007; Julian Barnes, *A History of the World in 10½ Chapters*, London, 1989.

Page 71. Vernacular names from Geoffrey Grigson, *The Englishman's Flora*, London, 1958.

Page 73. Midsummer fires: Christina Hole, *British Folk Customs*, London, 1976; Marcel de Cleene and Marie Claire Lejeune, *Compendium of Symbolic and Ritual Plants in Europe*, Ghent, 1999–2003; see also Grigson, *The Englishman's Flora*, op. cit.

Page 76. *The Leech Book of Bald* quoted in Eleanour Sinclair Rohde, *The Old English Herbals*, London 1972.

Page 78. John Aubrey, *Miscellanies*, 1890. 'Lay of the Nine Herbs', quoted in Rohde, *The Old English Herbals*, op. cit.

Page 79. Weed pheromones: V. S. Rao, *Principles of Weed Science*, 2nd edn, Enfield, N.H., 2000; Roger Cousens and Martin Mortimer, *Dynamics of Weed Populations*, Cambridge, 1995.

Page 81. William Turner, *A New Herball*, 1551. John Gilmour and Max Walters, *Wild Flowers*, 4th edn, London, 1969. Gerard, *Herball*, op. cit. Dodder pheromones: Consuelo de Moraes in *Science*, 29 Sept. 2006.

Page 86. Turner, *A New Herball*, op. cit.

Page 90. Charles E. Raven, *English Naturalists from Neckham to Ray*, Cambridge, 1947.

Page 92. Gerard, *Herball*, op. cit. (1633 edn).

Page 93. Thomas Johnson, *Botanical Journeys in Kent & Hampstead*, ed. J. S. L. Gilmour, Pittsburgh, 1972.

Page 94. Culpeper's life: Benjamin Woolley, *The Herbalist: Nicholas Culpeper and the Seventeenth-Century Struggle to Bring Medicine to the People*, London, 2004.

Page 107. Nicholas Culpeper, *The English Physitian*, 1652.

Page 111. 'Lively turning': Jonathan Bate, *The Genius of Shakespeare*, London 1998; see also Leo H. Grindon, *The Shakspere Flora*, 1883.

Page 116. John Clare's flower letters quoted in Margaret Grainger, *The Natural History Prose Writings of John Clare*, Oxford, 1983.

Page 118. John Clare, *The Shepherd's Calendar*, ed. Eric Robinson and Geoffrey Summerfield, Oxford, 1964. *Note:* There are no universally agreed versions of Clare's writings, given their unusual spellings and grammar (though the collected poems are published in their original orthography by the Clarendon Press, Oxford). I have quoted from a variety of more accessible sources, most frequently: *John Clare: Selected Poetry*, ed. Geoffrey Summerfield, London, 1990. This is the source of poems below where the title is quoted but no further reference is given.

Page 119. M. M. Mahood, *The Poet as Botanist*, Cambridge, 2008.

Page 120. Shepherd's-purse: 'The Flitting'. Daisy: *John Clare by Himself*, op. cit. Geoffrey Grigson quoted in Ronald Blythe, 'An inherited perspective', *From the Headlands*, London, 1982.

Page 121. Clare's journal note: *John Clare by Himself*, op. cit. Elizabeth Helsinger, 'Clare and the place of the peasant poet', *Critical Enquiry*, 13 (1987). 'The Lament of Swordy Well'.

Page 122. 'The Ragwort', *The Midsummer Cushion*, ed. Anne Tibble, Ashington, 1978.

Page 124. Enclosure of Helpston: Jonathan Bate, *John Clare: A Biography*, London, 2003; 'The Village Minstrel', *The Poems of John Clare*, ed. J. W. Tibble, London, 1935.

Page 125. 'Childhood'. 'Remembrances'.

Page 126. 'Cowper Green', Tibble, *Poems*, op. cit. W. R. Mead, *Pehr Kalm*, Aston Clinton, 2003.

Page 127. William Ellis, *The Practical Farmer, or the Hertfordshire Husbandman*, c. 1750.

Pages 130-31. Mead, *Pehr Kalm*, op. cit.

Page 132. George H. Whybrow, *The History of Berkhamsted Common*, n.d. (c. 1925).

Page 137. Kew: Lucile H. Brockway, *Science and Colonial Expansion: The Role of the British Royal Botanic Gardens*, New York, 1979; Alfred W. Crosby, *Ecological Imperialism: The Biological Expansion of Europe, 900-1900*. Cambridge, 1986.

# NOTES AND REFERENCES

Page 139. John Sibthorp, *Flora Oxoniensis*, 1794.

Page 140. George Claridge Druce, *The Flora of Oxfordshire*, 2nd edn, 1927. Grigson, *Gardenage*, op. cit.

Page 141. London ragwort: Rodney M. Burton, *Flora of the London Area*, London, 1983. Gallant-soldier in Malawi: Mabey, *Flora Britannica*, op. cit.

Page 143. Ruskin and ivy-leaved toadflax: Mahood, *The Poet as Botanist*, op. cit. Vickery, *Plant Lore*, op. cit.

Page 144. Thorwaldsen: Grigson, *Gardenage*, op. cit.

Pages 147–52. Quotes on weeds in America: Crosby, *Ecological Imperialism*, op. cit.

Page 155. Physico-theology: see, for instance, W. Derham, *Physico-theology; or a Demonstration of the Being and Attributes of God from His Works of Creation*, 1711–12.

Page 159. Burry Man: Mabey, *Flora Britannica*, op. cit. Peter Forbes, *The Gecko's Foot: Bio-inspiration – Engineered from Nature*, London, 2005.

Page 161. Ruskin, *Proserpina*, op. cit.

Page 165. Janet Malcolm, *Burdock*, New Haven, 2008.

Page 166. William Robinson's life: Mea Allen, *William Robinson 1838–1935: The Father of the English Flower Garden*, London, 1982.

Page 175. Pollan, *Second Nature*, op. cit.

Page 177. Robbins, *Lawn People*, op. cit.

Page 178. Henry D. Thoreau, *Walden; or, Life in the Woods*, 1854; *The Annotated Walden*, ed. Philip van Doren Stern, New York, 1970.

Page 181. Hemp in Norfolk: Michael Friend Serpell, *A History of the Lophams*, Chichester, 1980.

Page 192. Moths on dock. Chris Manley, *British Moths and Butterflies*, London, 2008.

Page 193. Wordsworth's poem: 'To the Small Celandine'.

Page 197. Gerard, *Herball*, op. cit.

Page 201. William Orpen, *An Onlooker in France, 1917–1919*, London, 1921. Captain Wilson's letter quoted in *War Letters of Fallen Englishmen*, ed. Laurence Housman, London, 1930.

Page 202. Trench gardening: Caroline Dakers, *The Countryside at War, 1914–18*, London, 1987. Ivar Campbell quoted in *War Letters of Fallen Englishmen*, op. cit. Ivor Gurney quoted in Dakers, *The Countryside at War*, op. cit. John Masefield, *Letters from the Front, 1915–1917*, ed. Peter Vansittart, London, 1984.

Page 203. Edmund Blunden, *Undertones of War*, 1927.

Page 204. *Country Life*, quoted in Dakers, *The Countryside at War*, op. cit.

Page 205. Clement Scott, *Poppy-land*, 1894.

Page 207. Macrae's letter: *War Letters of Fallen Englishmen*, op. cit.

Page 208. Poppy-day history: Mabey, *Flora Britannica*, op. cit.

Page 209. Robert Morison, *Historia*, 1680.

Page 210. Rosebay in Northumberland: George A. Swan, *Flora of Northumberland*, Newcastle, 1993.

Page 211. H. J. Riddelsdell et al., *Flora of Gloucestershire*, 1948.

Page 212. Salisbury, *Weeds and Aliens*, op. cit. Bucks rosebay story: Mabey, *Flora Britannica*, op. cit.

Page 213. Epigenetics: see, for instance, Jerry Fodor and Massimo Piattelli-Palmarini, *What Darwin Got Wrong*, London, 2010.

Page 214. BBC oral history website: www.bbc.co.uk/wwpeopleswar/stories. Rose Macaulay: Jane Emery, *Rose Macaulay: A Life*, London, 1991.

Page 216. Leo Mellor, 'Words from the bombsites: debris, modernism and literary salvage', *Critical Quarterly*, 46, 4 (2006).

Page 218. Ruins: Rose Macaulay, *Pleasures of Ruins*, London, 1953; Christopher Woodward, *In Ruins*, London, 2001; Uvedale Price, *Essays on the Picturesque*, 1794–8; Richard Deakin, *Flora of the Colosseum*, 1855.

Page 221. Bomb-crater flowers: J. E. Lousley, *Wild Flowers of Chalk and Limestone*, 2nd edn, London, 1969.

Page 230. William Robinson, *The Wild Garden*, 1870. For giant hogweed's early distribution: *Atlas of the British Flora*, ed. F. H. Perring and S. M. Walter, London, 1962.

Page 231. J. H. Dickson, *Wild Plants of Glasgow*, Aberdeen, 1991.

Page 234. Kenneth Grahame, *The Wind in the Willows*, 1908.

Page 235. Neal Ascherson, *Games with Shadows*, London, 1988.

Page 237. Alan Weisman, *The World Without Us*, New York, 2007.

Page 239. Detroit: Christopher Woodward, 'Nature in ruins', in *Urban Wildscapes*, ed. Anna Jorgensen (in press); Julien Temple, 'Last days', *Guardian*, 11 March 2010; also Temple's film, *Requiem for Detroit*, BBC TV 2010.

Page 240. Jonathan Silvertown, *Demons in Eden: The Paradox of Plant Diversity*, Chicago, 2005.

Page 243. Francis Lam: www.gourmet.com/food/2008/09/kudzu. Kudzu conspiracy theory: www.mindspring.com/~mdpas/kudzu. html.

Page 249. Mabey, *The Unofficial Countryside*, op. cit.

Page 250. History of tumbleweed in Dagenham: Burton, *Flora of the London Area*, op. cit.

Page 259. A. O. Hume, *Journal of Botany*, 1901.

Page 260. Anne Stevenson, 'Himalayan balsam', *Minute by Glass Minute*, Oxford, 1982. Dulverton balsam story: Mabey, *Flora Britannica*, op. cit.

Page 263. Sheffield miners' story: Mabey, *Flora Britannica*, op. cit.

Page 264. Japanese knotweed control: Lois Child and Max Wade, *The Japanese Knotweed Manual*, Chichester, 2000.

Page 266. David Pearman and Kevin Walker, 'Alien plants in Britain: a real or imagined problem?', *British Wildlife*, October 2009.

Page 271. Martin Sandford, *A Flora of Suffolk*, Ipswich, 2010.

Page 274. John Evelyn, *Sylva; or, a Discourse of Forest Trees*, 1664. Railway sycamore story: Mabey, *Flora Britannica*, op. cit.

Page 275. Ted Green, 'Is there a case for the Celtic Maple or the Scots Plane?', *British Wildlife*, February 2005.

Page 277. Peter Daniels, 'The Shoreditch Orchid', winner, Arvon International Poetry Competition, 2008.

Pages 277–80. Plotlands: Dennis Hardy and Colin Ward, *Arcadia for All: The Legacy of a Makeshift Landscape*, Nottingham, 2004; Deanna Walker, *Basildon Plotlands*, Chichester, 2001; Rodney L. Cole, *Natives and Aliens: The Wild Flowers and Trees of the Langdon Hills*, Basildon, 1996.

Page 281. Pesticide resistance: Cousens and Mortimer, *Dynamics of Weed Populations*, op. cit.

Page 282. Protected arable weeds: Watson and King, *Arable Plants*, op. cit.

Page 285. Jacques Nimki: www.nnfestival.org.uk/Contemporary-Art-Norwich, 2009.

Page 286. Tagworts: www.deptfordx.org.

Page 288. Olympic Park: Stephen Gill, *Archaeology in Reverse*, London, 2007.

Page 291. Carl Safina, *Eye of the Albatross: Visions of Hope and Survival*, New York, 2002

# Index